A Woman of Science

An Extraordinary Journey of Love, Discovery, and the Sex Life of Mushrooms

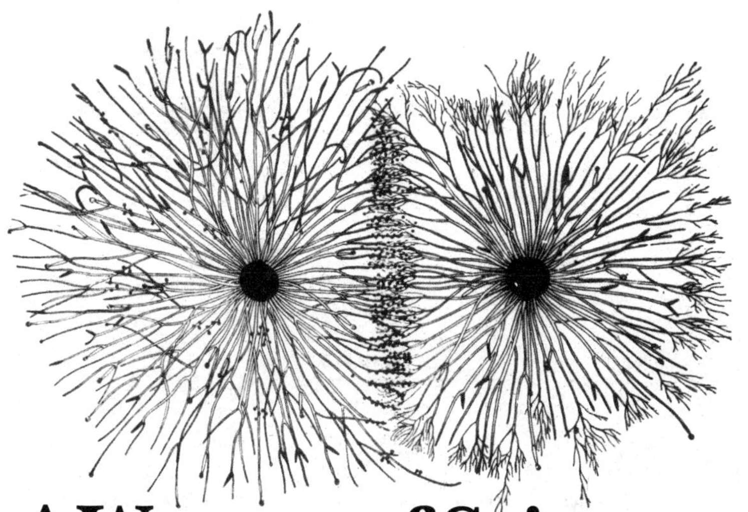

A Woman of Science

An Extraordinary Journey of Love, Discovery,

and the Sex Life of Mushrooms

CARDY RAPER

FOREWORD BY

REMELINE C. DAMASCO, MD

Hatherleigh Press is committed to preserving and protecting the natural resources of the earth. Environmentally responsible and sustainable practices are embraced within the company's mission statement.

Visit us at www.hatherleighpress.com and register online for free offers, discounts, special events, and more.

A Woman of Science

Library of Congress Cataloging-in-Publication Data is available.
ISBN: 978-1-57826-442-1

Design by Dede Cummings / DCDesigns

Printed in the United States

10 9 8 7 6 5 4 3 2 1

Advance Praise for A Woman of Science

WHAT MAKES THIS MEMOIR SO READABLE and engrossing is Cardy's frank account of her circumstances both in and out of the laboratory. Family photographs, and pictures and drawings of the organisms she worked with, add to the reader's interest.

This book encourages aspiring students—men as well as women—in how to overcome difficulties. Cardy Raper, a Harvard PhD, might well claim to have lived up to the slogan of the Harvard band, "*Illegitimum non carborundum: Don't let the bastards grind you down!*"

—PETER R. DAY, PhD
Professor Emeritus, Rutgers University
Author of *Plant-Fungal Pathogen Interaction*

* * *

CARDY RAPER SUCCEEDED IN BEING what she dreamed of as a young girl: a successful scientist with her own laboratory at the forefront of her subject. But it was not a conventional path to success. She married her teacher and mentor, a man much older than she, and devoted time to child rearing before she could return to science and work with him in his laboratory at Harvard University. John Raper died unexpectedly, leaving her a widow at 49. From then on it was a hard struggle. She had first to obtain her PhD and then find a job where she could keep her science alive. Eventually Cardy learned to be a molecular biologist, won independent research funding, and set up her own laboratory at the

University of Vermont. Cardy believed in herself despite the setbacks, a belief instilled early on by caring parents and five supportive older brothers. This is the personal story of an exceptional woman scientist.

—Professor Lorna A. Casselton, FRS
Foreign Secretary and Vice President,
The Royal Society,
Department of Plant Sciences,
University of Oxford

* * *

What forces attract a woman to science? What influences allow her to persist and ultimately thrive in domains that do not welcome female talent? This captivating memoir by Cardy Raper reveals the pulls and passions, as well as the setbacks, in the life within and outside the laboratory of one prominent woman scientist. As we learn about Dr. Raper and the people in her world, we also learn about the science of reproduction and the amazing capabilities of the many-sexed fungi. Both the human and the scientific stories convey one important lesson: If we are to thrive as organisms, we must prove adaptable in our methods but never stray far from our essential passions and purpose. I highly recommend this book to anyone interested in sex and gender.

—Faye J. Crosby, PhD
Professor of Psychology,
Provost Cowell College
University of California Santa Cruz
Author of *Affirmative Action is Dead; Long Live Affirmative Action*

CARDY RAPER SKILLFULLY INTERWEAVES THE scientific and the personal into a fast-moving and delightful read, offering candid and sage perspectives on a fascinating life.

—URSULA GOODENOUGH, PhD
Professor of Biology,
Washington University,
Author of *Sacred Depths of Nature*

* * *

CARDY RAPER DEMOLISHES THE CARICATURE many of us have of women scientists: nerdy, bespectacled, polysyllabic, introverted spinsters devoted to their (usually abstruse) field of inquiry. She's devoted, all right; not to just her science, but to life: From a childhood fascination with the natural world to a love affair with and marriage to a faculty mentor; from motherhood to widowhood; and finally from a voyage of discovery into what was at the time strictly a man's world, to worldwide recognition as a leader of research in the fascinating world of fungi.

Warmhearted yet vulnerable, often frustrated and saddened but never defeated, Cardy takes us with her as she enthusiastically probes the little-known world of fungi (where one species may have 20,000 genders), the often hostile atmosphere of male-dominated research, and the life of a working wife and mother.

—WILLEM LANGE
New England storyteller and
author of
Where Does the Wild Goose Go?,
Intermittent Bliss, and others

To the memory of John (Red) Raper who, as husband and mentor, showed me the way to a loving family life and the secrets of sex in the mushroom world. Remembrance of his luminous playful mind guides me still.

Contents

Foreword

THIS BOOK IS A STORY about the evolution of a modern woman trying to find her place. Women like me have benefited by women like Cardy Raper because of their accomplishments. It is refreshing to read her own struggles about where to take her life because it is something that we all can relate to.

I was captivated as Cardy discovers her passion for science, while studying volcanoes at the age of eight. She meets disappointment early in college where science is taught as facts to be memorized. Cardy transfers to another university and falls in love with her employer, and scientific mentor, Red Raper. We follow their marriage and family throughout the book, with one life-changing interruption: Red's premature death. While devastating at the time, Cardy slowly recasts this misfortune into the incentive to make it on her own in the male-dominated world of scientific research—a feat she might never have achieved had Red lived on.

At this point in the book, I was inspired to reflect upon my own career as a woman in the field of science and medicine. Cardy was truly a pioneer, and I am delighted to offer this brief Foreword to her remarkable achievement of a book.

Through flashbacks, readers experience the challenges

of a mother raising a son and daughter on her own, at a time when women working outside the home were not only rare but also taboo. Her courage and persistence are rewarded: she becomes a world-renowned scientist.

In *A Woman of Science,* readers discover something not well understood by nonscientists: how scientific knowledge is developed. This book conveys how scientists devise experiments to advance understanding, in this case about the myriad ways fungi accomplish sexual union. In language both accessible and passionate, Cardy shares her decades-long body of research, first on hormones governing courtship between male and female in a water mold, then on a mushroom-bearing fungus with over 20,000 sexes.

Her findings, with husband John Raper, lead to the astonishing revelation that molecules regulating mating and sexuality in fungi bear strong resemblance to molecules regulating development and sensory responses in many other organisms from flies, fish, and worms to mice, rats, and humans. This memoir describes not only the triumphs and dead ends of that research but also the difficulties of one woman's efforts to carry on alone as an independent scientist. Cardy Raper ultimately surpasses her husband's legacy, runs her own laboratory, and achieves her doctoral degree.

A Woman of Science also champions the relevance of curiosity-driven research to the understanding of fundamental biological principles, many of which are clearly applicable to humans. This book is a personal story as well, one I could not easily put down. I, too, celebrate the accomplishments of this inspiring scientist.

—Remeline C. Damasco, MD,
ABIM-Internal Medicine/Hospitalist
at Cheshire Medical Center Dartmouth Hitchcock
University of California of Davis, Medical School

A Woman of Science

An Extraordinary Journey of Love, Discovery, and the Sex Life of Mushrooms

Chapter One

Bereft

THE PHONE RANG AT 10:00 PM on Tuesday, May 21, 1974. A strained and unfamiliar voice announced the doctor's credentials: Resident at Peter Bent Brigham Hospital. Then,

"Is Mrs. Raper there?"

"Yes, this is she."

"Mrs. Raper—I'm sorry to have to inform you that your husband died a half-hour ago. . . .His lungs filled with fluid and his heart stopped beating. We did everything we could to save him. We worked for 20 minutes after his heart failed, but could not revive him. I'm very sorry. We tried our best."

I cannot recall my words of response. Whatever they were—maybe just a shaky "Oh no!" The phone dropped in its cradle. I stumbled back from the desk to our bed and gave in to deep sobbing. I could not bear the thought of telling anyone right then, not even our children.

Feeling the empty place on Red's side of the bed, I wept throughout the night trying to realize he would never be there beside me again. Never again could I stroke his silky skin, smell his pleasing scent, feel his warm caress, love his body whole.

By morning I recalled the days leading up to his last night. Red, as John Raper preferred to be called, suffered a spike of high blood pressure earlier that month and got slapped in the infirmary. His doc prescribed rest and some new medication. Red nonetheless insisted on finishing out the semester delivering his scheduled lectures in Harvard's introductory biology course, commuting from bed to lecture hall. Upon his release from "jail" as he put it, on Friday, May 17, we celebrated that fresh spring day by doing chores in the yard and enjoying an outside barbecue.

Suddenly, he couldn't eat the chicken. Shaking chills, then fever set in. He had recovered by morning, weak, but normal. Chills and fever returned that evening. This pattern continued with vengeance on Sunday: shakier chills, higher fever. I drove him to the infirmary. Numerous tests that night revealed no cause. He stayed there under doctors' care. Monday: no symptoms. By Tuesday noon, Red, seemingly recovered, anticipated his homecoming. In fact "the late Mrs. Raper," as he sometimes called me for my tardiness, arrived a half-hour late for a planned visit that day at 1 PM. He scolded as I entered his room and heard him complain about the lousy lunch he'd been served—signs of the Red Head getting better.

Then suddenly his chills returned, accompanied this time by dreaded symptoms of heart failure. I buzzed for the nurse, the doc, someone! No one came. His doc had cautioned earlier, "Should heart failure recur, you can decrease the burden on his heart by applying a tourniquet to the upper part of each leg." Frantically I grabbed the bed sheets and tied off both thighs.

Finally, a nurse, then the doctor arrived in alarm and arranged for Red's immediate transport to Peter Brent Brigham Hospital in Boston. As they wheeled his gurney onto the elevator, I pleaded to go along. "No. There is nothing you can do right now." Only *they* could be of use. The doc, the nurse, even Red agreed it would be best for me to gather his belongings, head on home, and visit the next morning.

I never saw Red again.

AFTER THAT FIRST PHASE of grieving, sufficient calm set in for me to start making the necessary calls. First to our children: Jonathan (senior at Harvard) and Linda (sophomore at Bennington); then the Departmental secretary, numerous Raper relatives, the Allens (my side of the family), and cherished friends. I seemed to muster enough stability to convey the sad news, without falling apart as I feared I would. The empathy, sadness, offers of help, overwhelmed.

Jonathan, the first to arrive, hopped on his bike and pedaled his way from Cambridge to Lexington. He brought comfort, telling of a visit he'd had with Red just the week before: "We talked about my plans to start graduate work in neuroscience at San Diego next fall. I told him how I wanted to go early and hike the Sierras ahead of time. Dad seemed really pleased. He loved those mountains and knew I would too. He even gave me his Elgin watch to take along."

Now on his way to establishing the kind of life Red wanted for him, Jonathan had reached a point of mutual respect and meaningful affection with his dad. True to self from boyhood to manhood, our son spoke with controlled emotion, eyes moist but not overflowing.

Then Linda hurried home. As daughter and mother, we held each other and wept uncontrollably—a needed release for us both. She grieved not only for our mutual loss but for the sense that her father would never be there to know what she would become. Our daughter had not yet found herself. She lacked assurance that her dad truly valued her potential. The hurt would dwell for years to come.

Stan Dick, Red's former grad student, arrived with a welcomed offer to help with a press release. My brother Luther and others all came by to do what they could do. I went into the lab at Harvard and tried to calm Red's students and colleagues who were in shock. I'd begun to realize just how much so many cared.

Now, the thought: *How can I go on without my buddy, the mentor who taught me to love the science of fungi, my lover, the father of our children not yet adults? We'd had the joy of working together, doing research in a lab he sustained. How can I do any of this alone?*

Chapter Two

Singleness

HALF MY SOUL HAD been ripped away. A life of loneliness loomed. But just after Red's death I had little time to grieve alone, except at night when sleep won over exhaustion. Red came to me in my dreams as a rather bedraggled wayfarer, who could not reveal the place he had been. Even so, I warmed from seeing him again. Waking disappointed.

In a letter to friends I wrote:

> . . . We take some comfort from the fact that ol' Red was his peppery self to the end—he complained of his dry, tasteless chicken for lunch at the infirmary, scolded me for getting there late that afternoon, chastised the nurses for not taking his temperature earlier, and, bless him, went down fighting. Why not? He did love life! And you know, in some ways, these last years were the best for him. He seemed in spirited good health since the scare about his heart two years ago, and—until this—quite confident he'd be okay for some time to come. We were looking forward mightily to the end of his tenure as Departmental Chairman and tackling some new research starting with a sabbatical term at Groningen, the Netherlands.

> There will be a memorial service for Red, Thursday next, May 30, at Harvard's Memorial Church, 4:30 P.M with Red's kind of music. Some of us will talk about the special meaning he had for us.

The service opened with a brilliant surprise that put me in tears of gratitude. Knowing of Red's and my love of trumpeting—we had both played trumpet in younger years—colleagues engaged the Cambridge Symphonic Brass Ensemble to play a Gabrieli Canzona from the back balcony. Then came a grand organ rendition of Bach's "Little" G Minor Fugue, recalling first flirtations when I, as a young student in Professor Raper's lab, whistled the opening canon one day—and he joined in with its second part.

I spoke first, noting among other things,

> There wasn't any sham about Red. If he liked something, he'd say it; if he didn't like something, he'd let on about that. . . . As parents we complemented one another. Along with the love and caring, Red provided a good dose of discipline. He claimed regret that I would not allow him to discipline the kids the way he disciplined the dog. He was the one who always fixed things for them, answered their hard questions, got them to think and do for themselves. . . .
>
> He had a fine life. I wish there'd been more. The pain of losing him is great. But as I feel sadness, a feeling of gladness comes in. . . . His life left ours richer. All the things he could do—and there were many—he tried to teach us to do. I think I'm a lot better off from his tutelage: I can change a tire, shingle a roof, run a drill press, handle a power saw—he was the one who helped me most to understand the joys of research and to know how to do it.

Others spoke: Red's brothers Arthur and Ken, a student, and several colleagues with whom Red had taught. The service concluded with a moving performance of parts from Bach's B Minor Mass, sung by my own choral group behind the chapel screen.

A hundred or more letters from relatives, friends, students, colleagues, Bio Lab secretaries, and workmen poured in. They encompassed a tremendous expression of energy—an almost unbearable sense of emotion. Each communicated some unique feeling about that person's special relationship to Red and to me.

One friend, Alvin Levin, a local lawyer and paraplegic who knew suffering firsthand, remembered Red in a powerful way:

> It is unpardonable that Red's vitality and enthusiasm be ended—and as I write that, I hear him, pursuing even his most pessimistic judgments with enormous gusto and energy. The two of you brought us so much joy. That must have been because Red and you knew the secret of getting the very sap out of life and savoring it fully . . . Hold to that, dear Cardy. The pain and loss are damn bad . . . but in your coming together with Red you have made ripples in the world which could outlast the relics of the Romans . . .
>
> *Wow!*

Gratifying words came from friend and colleague Ginny Page who knew " . . . what it's like to have half of your being torn from you forever . . . " since the untimely death of her husband Bob. "I hope you will take some comfort," she wrote, "in the fact that some of us who knew Red long before you did felt you were the best thing that ever happened to him."

Alvin Levin's wife Betty sent some wise advice about coping:

> The service this afternoon reflected you and Red entirely and was a sustained and sustaining experience of simplicity and beauty . . . most of all, I appreciated and loved what you said and the way you said it. When people speak

of courage, I think they tend to miss the point. You do what you do because you are who you are, because you have to be you.

But there's a hitch to managing as you do and that's what I call the rake effect: You know, when you step on it and the handle flies up and hits you in the face. You think you know it's there, you're watching out for it, and it happens anyway. Always when you're off guard. With strong people who cope and bear their grief, doing all the things they do, getting used to it—they think—the system takes its toll. . . . I hope this summer will give you a chance to breathe, to weep, and, if you're unable to duck when the rake handle flies up, at least a chance to stop because it hurts so damn much.

Betty was right. I first *stepped on the rake* listening to a passionate rendition of preludes and fugues from Bach's Well Tempered Clavier at a friend's home in Lincoln, Massachusetts one Sunday afternoon. The tears flowed uncontrollably.

Now 49, I dwelt on the past and thought of my future as widow and mother—my cherished life in science seriously in peril. With an appointment officially termed "Technical Associate" and only a master's degree, I lacked the formal credentials to establish an independent position and carry on the work Red and I had accomplished together in his lab.

My world as I knew it had crashed.

Chapter Three

Seeds of Encouragement

SOME LETTERS OF CONDOLENCE from former associates and students encouraged me to continue with the work Red and I had shared.

A former grad student, Judy Stamberg, then on the faculty of Tel Aviv University, wrote,

> We would like you to think very seriously about spending some time in Israel in the near future. The bi-national project we discussed last summer, with Red as co-sponsor, is about to get underway and we would very much like your collaboration. Your scientific career certainly shouldn't terminate now—as a tribute to Red and because you can contribute a lot.

The "we" Judy referred to was Yigal Koltin, a fellow faculty member at Tel Aviv, and a former student and postdoc in Red's lab. He wrote,

> . . . It is my feeling that you can and should continue your scientific contribution. You certainly have much to offer. I imagine that it is difficult to leave the kids at such a time, but I would strongly encourage you to go to

> Holland. Furthermore, if you would like to join us in our research for 6 months, we would be more than happy to have you come as the collaborator.

I was working in the lab when Judy and Yigal were also there a few years earlier. I knew they valued my skills back then, but their confidence in my ability to carry on now provided a much-needed sense of hope.

Within a week, things began to fall in place. I received a surprising call from Lawry Bogorad, the newly appointed replacement as Chair of Harvard's Biology Department: "Cardy, the Department met and agreed to pursue an appointment for you as Research Associate. We know you played a big part in Red's research and think you should go on with it. Are you interested?"

"Well, gosh, yes. I am. But I don't have the doctorate. How can I meet the requirements?"

"You'll have to assemble the credentials you already have: curriculum vitae, letters from established scientists, maybe apply for transference of the Campbell Soup Company grant from Red to you to continue research on the edible mushroom."

I scrambled. The appointment came through within two weeks. It included the privileges of a lectureship in microbiology and ample laboratory space to continue research despite my lack of the usual PhD credential. I was amazed and gratified. *How can such a venerable institution accommodate me, a woman with only a master's degree, as an independent scientist? I hadn't even dreamed of it. Well, it happened. I'll go for it!*

The future seemed less bleak—at least for now.

BY THEN IT WAS June, the time Red and I had planned to head for Europe and start his sabbatical leave with Professor Jos Wessels at the University of Groningen, Haren. We had already ordered two airline tickets to Göteborg, Sweden, and a Volvo to be picked up at the factory. Daughter Linda and I decided to go ahead with the plans and spend the summer traveling, heeding Yigal's advice to include a visit to Wessels' lab in Holland.

Upon arrival in Sweden, we met our new ride and a colleague from Finland to enjoy a brief holiday at a nearby lake. We then headed south.

The joy of owning a spanking new Volvo began to wane somewhere in Germany when its temperature gauge registered red and a cloud of steam spewed from its hood. No German mechanic seemed able to fix the problem. We limped to our destination in Haren where Professor Wessels, muttering anti-German epithets, steered us to a reliable Dutch mechanic who fixed some fault in the alternator. Oh well, the trip so far had fulfilled its purpose as a diversionary expedition.

Haren's charming thatched-roof homes lining pristine streets seemed the scene of some fairytale. Jos and his family took us in and showed us around. The University, with its modern buildings and exotic Botanical Garden, impressed me as a place to spend quality time. The reality of Red's recent death hit hardest when I saw the delightful little townhouse they had chosen for Red and me to lease during his sabbatical year.

One of Wessels' students had just passed his doctoral defense, an elaborate affair in which the candidate, colorfully robed, sits in front of an invited audience and answers questions posed by he himself as addendum to his thesis. Not only that, but upon acceptance, the newly minted Doctor of Philosophy stages his own celebratory party at which *he* treats his committee of professorial judges, friends, and relatives to wine, beer, and *snacken.* In recounting this protocol, Wessels surprised me by suggesting that I consider joining his group as a doctoral candidate under his aegis—an attractive scheme I then thought outlandish, but it planted a seed for the future. I wondered, however, whether Wessels encouraged me for my own merits or as a tribute to Red's wife. *I shall ponder that later.*

My brother Luther, Professor of Political Science at the

University of Massachusetts, was spending that summer in France. He'd picked up a Peugeot and met us in Amsterdam, where Linda and I shipped our now friendlier Volvo off to the States by slow freight. Lute aimed to study a bunch of newly planned towns in southern France. Linda and I went along for the ride: lazy stays at small country inns, succulent picnics in shaded olive groves, sleepiness in sleep-worn beds. With occasional lapses into waves of grief that struck without warning as the *rake handle effect*, this journey marked the beginning of a long period of healing.

By the end of July it was time to head home and try to get on with my lifelong passion for science.

Chapter Four

Volcanic Beginnings

I WANTED TO GROW up and *be* a scientist ever since third grade, 1933, when Dr. Rusterholtz, Professor of Education at the local teachers' college, came to our class once a week and taught science as grand adventure. In one session, he had us all using heads and hands to think about volcanoes: With flailing arms, he talked of powerful eruptions that spew hot lava from deep in the earth—a concept utterly foreign to those of us born and brought up in northeastern New York State, where nature's most dramatic events are frozen lakes, blizzard whiteouts, and rare displays of aurora borealis. Dr. Rusterholtz not only volubly described volcanoes; he enticed us to model them.

Griffie Larkin and I chose modeling clay. We built a mountain with a hole down its center, fixed a small fan and an upright stick at the base of the hole, attached a strip of red cloth at the top, and rigged an on-off switch to the fan cord extending from the bottom of the mountain. Hey! Click the fan on, and the rippling red cloth simulates eruption.

"Wow, it has action! Neat!" said a classmate. Griffie and I thought so too, though we envied my best friend Anne and her partner Phil for making fake smoke with dry ice—and having the foresight to get some.

A first awakening: *Science is action packed! Not just the passive business of memorizing facts to ace a test.*

One evening in early June, Dr. Rusterholtz took me, my brother Jonny, and some of our classmates on a supper hike to the bare top of a small Adirondack mountain overlooking Lake Champlain. Licking lollipops for dessert, we lay on our backs gazing up at the darkening sky, each of us hoping to be the first to spot a star. Jonny yelled, "I see one! I see one! Over there, just over that mountain with the big rock slide!" Then we all saw a bright pinpoint of light in the salmon-pink hue of sunset. Gradually more and more stars appeared, until a great white streak of them glowed high overhead.

Dr. Rusterholtz talked about the science of the night sky and how astronomers know that the brightest steadiest lights come from planets like our earth, revolving about the sun, reflecting its light. The fainter twinkling lights are other suns far away, and the massive streak of lights are the nearest stars of our own galaxy, just one of billions in the universe.

"How did those scientists find that out?" I asked. "Where did this universe come from anyway? Is it changing all the time? Will our sun last forever?"

"We don't know all those answers yet," our teacher said, "but astronomers have found a lot of clues that predict our sun will burn up and burn out billions of years from now. Beyond that, new discoveries are made every day, and a lot more will be known by the time you kids are grown. Maybe even one of you will find out something new about the universe some day."

Ah, science can be a journey of discovery!

Jonny and I got so inspired I announced at the family dinner table the next day, "When we grow up *we* want to be *scientists*!" Mom responded, "That's nice dear; Jonathan can be a doctor and you can be a nurse."

"But, Mom, I don't want to be a nurse. A nurse is *not* a scientist!"

This was the first of many obstacles I would have to overcome in my lifelong ambition to become a scientist—maybe not an astronomer or geologist, but some kind of discoverer. Mom just did not understand. How could she? In her world, scientists were medical practitioners, a hierarchy of men supported by women. Are women not allowed to be in charge? There were no obvious role models.

My mom's mother had died leaving five children when Mom was only seven.

As the only two girls in the family, she and her sister had to take on some household chores, including the Saturday morning ritual of baking a week's worth of bread. She told of her two older brothers, coming in from their chores in the field, snatching the loaves, and tossing them back and forth as footballs. The boys giggled; the girls cried.

When I complained about my own brothers' teasing, Mom had the wisdom to advise: "They just do that to get your goat. Either tease them back or don't pay any attention

My motherless mom in the middle

at all." Well I had five older brothers, not just two, and they all seemed to rule the roost.

When a very little girl, I warned *them*, "You just wait 'til I get to be a boy!"

Even so, I always knew they loved me, and I loved them. I just wanted to have the advantages they had.

As my mom approached her teenage years without a mother of her own, she had scant knowledge of the physiological changes an adolescent experiences. Neither her father nor local aunts helped in any manner. It was thought improper to speak of such matters in that late Victorian age.

When I became a teenager, Mom gave me a book about birds and bees. As we segued uncomfortably to the human condition, she imparted the essential, timely knowledge she herself had never received.

My mom came a long way from most women of the preceding generation. Besides fulfilling the usual role of wife and mother, she had certified at the local teachers' college as an elementary school teacher, taught on Long Island for several years, and continuously worked toward civic improvements. She did not, however, go so far as to march with suffragettes; when old enough to feel outraged at the very idea of her being denied the right to vote, I asked why *she* didn't march. She answered, "Ben (her husband and my father) wouldn't have liked it." I hardly understood how this forceful woman could have been so acquiescent.

Nevertheless, Mom gave strong voice to political thoughts, and once given the right to vote, she voted every time. Unlike most women then, she fought for causes: education, libraries, the rebuilding of our burned-out church. While fulfilling the traditional role of woman in her time, she had taken a step beyond the women of her mother's time. I admired her convictions. She expressed them in the only way acceptable for a female of the early 20th century.

Both she and Dad (a farm boy who became a small-town lawyer) cherished and loved me as their only daughter. They wanted me to flourish, be educated, marry an educated pro-

Five boys and one girl

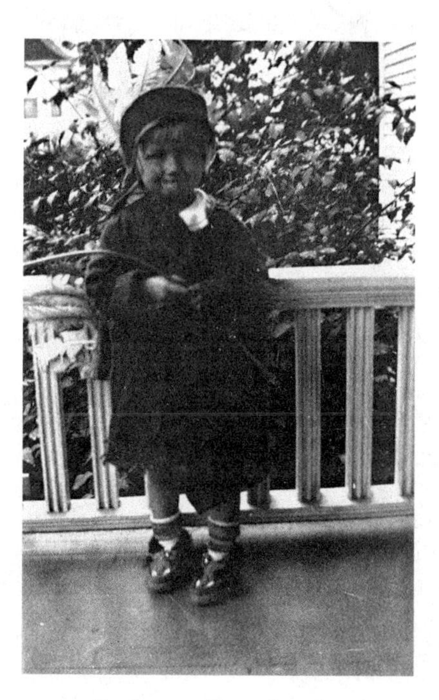

Cardy wanting to be a boy

My youthful mom Nell

My youthful dad Ben

fessional man, then, as homemaker, raise educated children. They sought education and marriage for their five sons also, but a professional life as well. The idea of *my* becoming a scientist seemed beyond their ken. Silly me, I dreamed of celebrity, being at least some kind of heroine. I wanted to do what boys did. I wanted to ski, play tennis, become a leader—be valued for who I was, not just for looking female pretty. I wanted that too, of course. I liked flirting with boys, but they had to be interesting. And they had to respect me for what I could do.

As I grew up in Plattsburgh, New York, a clear sense of gender bias prevailed all through high school. Even though the two strongest, most dedicated educators were female and college educated, they encouraged boys more than girls to go on to college. Gracie Barker (the Principal) and her sister Gertie (our Latin teacher) had graduated from Wellesley College at the end of the 19th century. Both sisters—we called them the "Barkae"—might have been expected to nurture girls more than boys, but they didn't. Neither was married. Maybe they thought too much education would spoil a girl's chances for wifehood.

WHILE MY FOLKS WANTED wifehood for me, they also urged me on to college. I even got a scholarship for Syracuse University.

But disenchantment marked my first years there where science then was taught as a body of facts to be absorbed and regurgitated in written tests. I had loved the physics taught by Mr. Benton in high school. He conveyed the logic of a concept by taking time to develop and illustrate the relevant principles—the way a fulcrum works, for example, as the pivoting point of a lever—a fascinating exercise that made his stale jokes tolerable throughout the entire year. In college, however, I had to memorize formulae with no time allowed for thoughtful derivation. For botany we learned the Latin names of trees, shrubs, and plants without understanding relationships between form and function. Microbiology dealt a similar litany of facts. I rebelled so strongly I began to consider switching to either another field or another university.

While professors in other subjects piqued my interest in sociology, music, and political science, several aspects of campus life left me dismayed. The majority of female students seemed focused on finding a mate among the transient young officers-in-training on campus or the few male students not yet called into military service. Football and partying took precedence over learning and scholarship.

No decent dorms existed then. Victorian homes converted to campus cottages provided the only non-fraternity accommodations. In search of more comfortable living space, I joined a sorority in my sophomore year only to learn later that I could not abide its snootiness and intractability toward pledging Jews, Catholics, or anyone with a drop of Negro blood. A worldwide war was going on. Four of my brothers were in it. Yet most fellow students, including my sorority sisters, sought trivia, fun, and romance as the foremost objectives of college life.

Three of my sorority sisters, Allie, Ole, and Conant, shared my views. Sweet, conscientious, cautious Allie roomed with me my sophomore year. We shared a fondness for each other but parted ways when I "borrowed" her shampoo once too often. Ole, a more forgiving sort, agreed to be my roomie during junior year. She became the soul mate I had wanted ever since high school. An artist with an exotic full face and hints of Asian features, Ole was the only child of Norwegian immigrants. (Her father, Ivan, was one of those legendary glass blowers at Corning Glassworks.) I adored Ole's sense of artistry and aesthetics. We both loved classical music and connected strongly in our appreciation of the absurd. I loved her willingness to take risks and seek adventure. Sometimes we dressed like boys, tucked our tresses under our caps and hopped a city bus to the end of the line, just to experience new neighborhoods—an adventure thought too risky for a couple of young girls at that time.

Then, Conant, a bookworm, satisfied my tomboy instincts. We shared laughs over childhood pranks we had played with our respective brothers and their gangs. Conant always had a story to tell. I thought she would grow up to be a writer, but she didn't. Savvy in her ways, however, she helped me when I needed advice about an unwanted pregnancy some two decades later.

We four rebels talked endlessly of social injustices and how we might amend them. Alas, attempts to reform our sorority sisters failed miserably, due no doubt to *our* lack of subtlety in conveying the degree to which *they* were misguided. While the sorority imbued social skills such as "circulating" at parties, other precepts struck us rebels as downright unprincipled. I recall a hopeless confrontation in the laundry room one day with older sorority sister Mary Lou. As we hung undies and ironed blouses, I vented my outrage at a recently issued dictum designed to increase Big Women on Campus (BWOCs) from within our sorority: "You are to join as many organizations as possible and achieve high office whether you're interested or not, so we can score big in the next round of competitive pledging!" Mary Lou turned and paused at my ranting. She stared, eyes glazed, like an immobile cow—and said nothing.

Gatherings offered delightful diversion. I recall a sorority sister's wedding at her mother's home nearby. The warmth of that spring day generated thirst; a large bowl of champagne punch seemed innocently tempting. I guzzled several cupfuls. When I went to pee, the pattern of tiny black and white tiles jumped up and down on the bathroom floor. I knew then what Eleanor Roosevelt meant in her speech on campus last week when, in her high-pitched voice, she exclaimed, "College is a learning experience in which every young girl should know *just* how much she can drink." Eleanor was my first female role model—her wisdom and social activism, first as First Lady and later as Franklin's widow, has guided me since early adulthood. My mother, on the other hand, thought she should have spent more time looking after her six children.

Thoughts of switching from science to another discipline—perhaps sociology or political science—began to emerge in a series of bull sessions in the smoker room where sisters sucked on cigarettes and sometimes took up topics beyond who's going with who. We rebels, Ole, Allie, Conant, and I, joined the smokers because it seemed the cool thing to do. I guess we aimed to learn something and perhaps persuade others to share our views.

We pondered, for instance, the reasons for war. Could it be *ethnocentricity*, the tendency to view other races and cultures by the criteria of one's own race and culture; to think, *my ways are right; other ways are wrong*? We had acquired this concept from an excellent course in cultural anthropology taught by Professor Herring (affectionately known as "Hose Nose Herring"). He proclaimed the dangers of this all too human trait and how it could be an important cause of conflict, from family feuds and tribal warfare to wars between nations. We pondered how *we* might overcome such attitudes of privilege and superiority even in our midst. Of course no one had answers.

Engaging as they might have been, those sessions initiated a lasting penalty: the misery of getting hooked on cigarettes and going through hell to quit two decades later, when scientific studies showed what my mother always knew: "Smoking is like putting nails in your coffin!"

A particularly bitter memory remains: During our noontime meal, April 12, 1945, someone came to the sorority's dining room door and announced the sudden death of President Franklin D. Roosevelt. Except for Ole, Conant, Allie, and me, everyone looked up at this brief interruption, then went on eating as if such news meant little. We misfits stood in shock, left the room, trudged up the hill to the drafty Crouse bell tower, and signaled the campus at large. Leather straps connected the bells to an ancient set of wooden levers. We punched those levers as never before, playing the dark opening theme of the Adagio from Beethoven's Eroica Symphony over and over as a campus-wide tribute to our heroic leader, who had brought us through the Great Depression and much of the Second World War.

That did it! Fed up with failed efforts to effect reform, Ole and I devised a plan to abandon a hopeless cause by transferring to another university. In the midst of our junior year, we felt held back.

Ole did the research and identified the University of Chicago as the opposite of Syracuse, surely more in tune with our values. Under the leadership of Robert Maynard Hutchins and his colleague Mortimer Adler, the U. of C., with its Great Books Program of undergraduate studies, was famed as a citadel of scholarly pursuit. Lacking the rah-rah spirit of major varsity sports, it supported not even a football team, its athletic stadium having been usurped by the U. S. government to serve as a site for accomplishing the first chain reaction leading to development of the atomic bomb.

The idea of switching universities attracted Conant and Allie, but they simply could not commit to such radical change. Ole and I made up our minds: If our parents wouldn't sanction the transfer, or Chicago wouldn't accept our credentials, we'd join the Women's Army Corps (WACs)—so there!

We never had to join the WACs. Our parents agreed to the transfer without a whimper; the University of Chicago took us without loss of credits. After three years of growing disenchantment, I had chosen to change my place of learning but stay with my first love: science.

Except for Conant and Allie, the majority of sisters were doubtless relieved to see Ole and me leave, taking our phonograph and loudly played records of Bach, Beethoven, and Bartok with us. They doutless could not understand how we could abandon the comfort of a pleasant sorority home for the big bad city of Chicago, just for academic reasons.

It was a move of significant consequence.

Chapter Five

Water Dalliance

THERE I WAS AT 21, a biology major in my senior year at the University of Chicago and not yet a scientist. My thoughts spun: *With a bachelor's degree I could be a lab tech, or perhaps a teacher in some junior college, but not my own boss making new discoveries. How do I get from here to there? Perhaps graduate school is the way to go, though I've not been encouraged to have such ambitions through all the schooling to date. Would my folks agree? They've already put five boys through college and beyond; surely finances are scarce.*

A solution to my puzzlements began to materialize through a friendship I'd established with a lean lanky communist sympathizer called Lou. I'd met Lou while auditing a course on international relations. He possessed a sharp brain full of esoteric facts that neatly fit his bias toward communism as the best form of government yet conceived. I, a budding scientist, rather enjoyed confronting his strong tendencies to reason subjectively. While we both liked the sparring, Lou professed a physical attraction to my long blond hair, bright blue eyes, rather large nose, and slender female frame.

Then, even more than now, the way a female looked held sway. Alas, I could not reciprocate—Lou's physiognomy held no appeal; nor did his non-scientific way of reasoning.

Still, Lou did have my interests at heart. He wanted me to consider graduate school. He urged a meeting with an old friend who would be starting as Assistant Professor of Botany the summer following my graduation. The friend, John Raper, was about to transfer from his position as radiobiologist at the Manhattan Project in Oak Ridge, Tennessee, to the University of Chicago. He sought someone interested in doing graduate work on fungi. Lou hoped I would qualify. He arranged a meeting.

Raper arrived on a brisk sunny day in April. The three of us met at the corner of Ellis Avenue and 50th Street. My reaction at first meeting: *This wiry, sharp-faced redhead in rumpled clothing is a university professor? A fungal expert? Prefers the nickname "Red"? He looks like a college sophomore, although Lou said he's 34.* His looks bore a resemblance to no one I'd seen before, with the possible exception of the movie star Rex Harrison. He seemed forlorn somehow, yet brash. *Damn,* I thought, *uncommonly attractive.*

We had lunch at a "greasy spoon" on 55th street within a few blocks of campus. I can't recall a thing eaten except its greasiness. Animated conversation charged the air. Lou and Red had met at Cal Tech just before the war and had not seen each other since. They played catch-up. I listened with interest, responding to the occasional query cast my way.

So this was meant as an interview for a prospective lab tech and grad student? Well, the guy seems fascinating, fiery brilliant, full of life. I think he liked me from the start, though his darting gray eyes seldom fixed on mine. On the walk back to campus, I noticed his flaming red hair bouncing crazily as he outpaced both Lou and me. We parted with little formality at the street corner of our meeting.

Back home alone in my cozy rented room with its pink chintz window treatment, matching chair covers, and ugh-green carpet, my thoughts spun: *What if I don't get the job? Here I am, about to graduate without another prospect. Do I really want to do this anyway? Lou thinks so. He wants me to stay, thinking there is yet a chance I'll embrace communism and become his wife. This*

Cardy, wanting to be a scientist, in love

Red

intriguing friend of his would likely not compete, being 13 years older, safely married, and the father of a newly adopted baby boy. But I needed something beyond Lou.

LESS THAN A MONTH after meeting Red Raper, I returned from class to my room to find a large package with a chewed-off edge sitting by the door. It contained a stack of papers by John "Red" Raper. Each had a bit of its lower right-hand corner missing as if some mouse had been at it. I read as much as had remained intact with utter fascination—all published accounts of Red's prewar research on a little water mold called *Achlya* (pronounced ah-klee-a), a sexy fungus that signals its gender via an array of endogenous hormones (now better described as pheromones).

Through a series of cleverly crafted experiments, Red had shown this small critter to be an opportunist: it can become either male or female depending upon the partner it meets. If it encounters a mate more male than itself, it acts as female. A mate more female than itself elicits maleness. It swims about in freshwater ponds as tiny spores propelled by the whipping motion of two fine appendages called flagella. When a spore settles down to take in nourishment, it develops into a fuzzy ball of mycelium made up of many long threadlike cells called hyphae.

Red had proven the unorthodox lifestyle of this lowly water mold by microscopically examining cultures collected from nature. He detected sex signals (later identified as steroids similar to human sex hormones) by interposing a water-permeable barrier between isolates and observing sex organ development as the isolates tried to make contact.

A strong female announces her presence by continually releasing a substance Red called *Hormone A*. If the nearest fuzzball happens to be relatively male, it responds by differentiating special rod-like male organs called antheridia. Then, and only then, does the male produce *Hormone B*, causing the

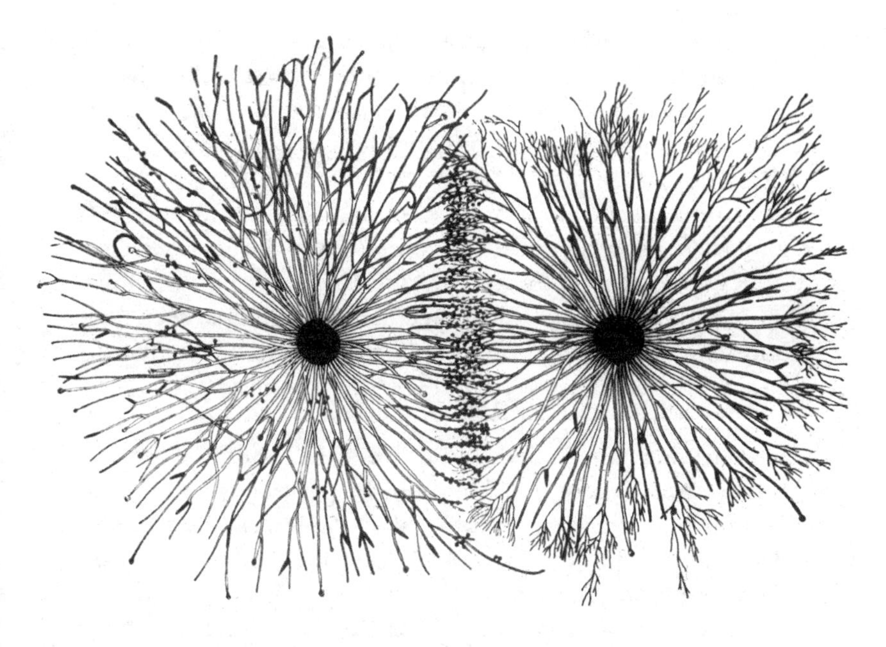

Red's hand-drawn illustration of Achlya *mating*

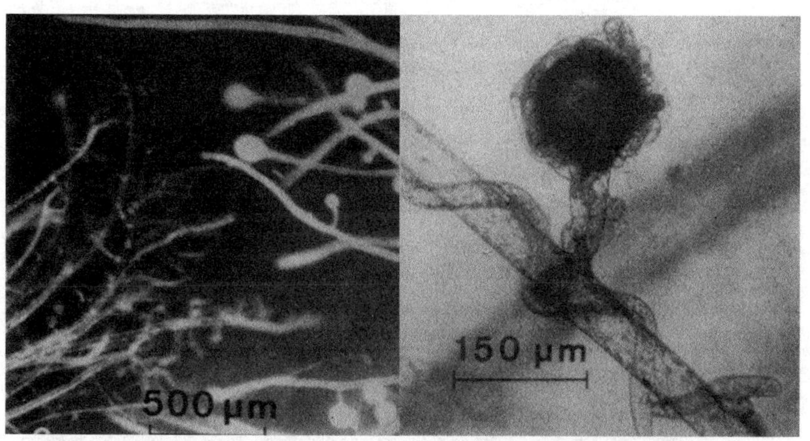

Female on right attracting male on left

Male wraps around female attempting fertilization

female to make her sex organs called oogonia—stalks bearing little round egg-containing structures resembling lumpy lollipops. The lollipops produce yet another substance, called *Hormone C,* attracting the long tubular male organs all the way to attempted contact.

In the absence of a barrier, antheridia fuse with oogonia and discharge their fertilizing contents into the enclosed eggs.

Raper extended proof through other ingenious experiments. In one example, he detected sex signals between male and female by placing them separately in connecting micro-aquaria he had designed and constructed himself.

Bizarrely, an individual can mate simultaneously as female on one side and male on the other, depending upon the relative maleness or femaleness of the partner on either side. Hmmm, an amazing creature that does not waste energy developing into one sex or the other until the time is ripe. Female speaks first to male, male answers female, then female beckons—all from a distance. As Red exclaimed in a later publication, "Thus in *Achlya,* as perhaps in other, even higher forms of life, the female is the initiator in dalliance."

Aquariatized Achlya

I was so impressed with the cleverness of these experiments and the elegant manner in which they were described, I decided then, without doubts, that Red was the scientist I wanted as mentor. Never before had I become acquainted with such an imaginative and comprehensive approach to discovery in the laboratory.

But why should I care so much about how *Achlya* does sex? Was I envious or just plain curious? Did I admire the opportunistic exquisiteness of a system in which female hails male and male responds by telling her, "Get ready; here I come"? Then, female says, "I have formed my receptacle in response to your emanation; come hither—we shall meet and test our fitness!" Why and how did this scheme develop in the lowly water mold and then get lost throughout evolution? Or did it get lost?

Of all the men I have met, few made me tingle and want to embrace. Are we humans attracted through fumes from a distance? Might there be some common threads to the sex life of water molds and humans? These are questions I wanted answered.

Curiosity about both *Achlya* and Red Raper strengthened my will to apply to the graduate program in the Department of Botany—pending, of course, my parents' permission.

A CALL TO RED confirmed the likelihood that some cute little field mouse at the country place where he was staying that summer had chewed off the corner of his package to me while it sat on the porch awaiting mail pickup. We talked about the packet's contents: his research papers. My enthusiasm for his work must have conveyed assurance; I am not at all sure he had other choices. In any case, he offered the job as research tech at $100 a month and agreed to take me on as a grad student working toward a master's degree in Science, pending acceptance of my credentials by the Department of Botany.

A call to my parents shortly thereafter met with surprisingly little resistance. They had never encouraged or expected me to go beyond a bachelor's degree. I guess Pop's small-town law practice was doing okay, now that the war was over. And they liked the idea that I could pay halfway. Perhaps their consent had something to do with my not having yet found a husband?

Chapter Six

Mentoring

THE CONVOLUTED PATH LEADING Red as mentor to me developed over a decade earlier with his rescue from delinquency while an undergraduate at the University of North Carolina. As the youngest of eight siblings, born and bred on a poor tobacco farm outside Winston-Salem, Red craved independence. Like his brothers, he wanted to leave the tedium of farm life and do something else, anything but farming. Red's brother Ken, three years ahead at the same University, sensed disaster. He saw Red floundering, going to parties, playing his trumpet half drunk at midnight on the roof of his dorm, getting poor grades. Ken offered a bit of counseling. "John," he said (Red was always John to Ken), "Why don't you sign up for John Couch's course in mycology next term? Professor Couch is a ball of fire, and he'll teach you about some amazing creatures." That advice saved Red from bad ways. Couch turned him on to fungi, just as Ken thought he might. With John Couch as mentor, Red discovered *Achlya* and inklings of its sex life.

Red followed Ken to Harvard, where he carried on the *Achlya* work under William H. ("Cap") Weston, who spawned a myriad of fungal experimentalists. With humor, theatrics, and know-how, Cap mentored no fewer than 28 graduate students and postdocs (all male, of course), who then went on to become the leading experimental mycologists of the age. This first generation spawned a second generation of 83 mycologists who, in turn, inspired another 48, who then begat a fourth generation of at least 33 by the year 1990. I wonder, in retrospect, could such prolific spawning—primarily of men— explain my difficulties of finding a job as a woman in that crowded field several decades later?

Red's years at Harvard shaped him in many dimensions: He gained fame for his research with *Achlya*, persuaded his North Carolina sweetheart Ruth Sholz to join him in marriage and, upon becoming Dr., won a National Research Fellowship allowing him to go wherever he chose for an additional year of study. Red chose the lab of well-known plant biologist, John Bonner at Cal Tech.

Had Couch and Weston not encouraged Red to trust his own instincts, he might have ended up studying bean seedlings instead of *Achlya*, and Red might never have become *my* mentor. When he reached John Bonner's lab at Cal Tech, Bonner offered a firm greeting: "So, Raper, you want to work on my bean seedlings, do you?" "Hell no!" said Red. "I came here to purify *Achlya*'s *Hormone A*."

Neither gave in to the other, so Red marched down the hall and entered the lab of Dr. A. J. Haagen-Smit, a biochemist who agreed to teach him how to chemically fractionate this sexual hormone with an aim toward characterizing its properties.

Thus began what became a project for me some eight years later.

Chapter Seven

Graduate Studies

WHILE AT CAL TECH, Red grew about 400 gallons worth of female fuzzballs to extract a minuscule two milligrams of a partially crystallized creamy white concentrate of *Hormone A*. Enriched by 70,000 times of the starting material, it could induce male organs when diluted one to ten trillion. Red defined certain properties of the hormone, but its high activity prevented definitive analysis, given the instrumentation available at that time. In any case, the power of this concentrate to create a measurable developmental event—think development of a male penis—gained attention in the world of biology.

Red continued with his studies of *Achlya* as beginning professor at Indiana University the following year, but he had to suspend it during the war, when he served as radiobiologist for the federal atomic energy commission known as The Manhattan Project at Oak Ridge, Tennessee. Most of his precious strains got lost in storage. A Caltech friend left in charge, Harvey Tomlin, conveyed the sad news in a letter:

> The *Achlya* project has been root-balled around considerably . . . Its care was turned over to another party who messed around with the project for nearly five months

> and then went over to Chemistry. The care of the little beasts was left in the hands of a helper he had. I [tried to] start them all [up] again, and this was the beginning of my chills and fever, for upon examining the cultures, I found them [almost all] contaminated.

When Red resumed his *Achlya* studies at the University of Chicago in 1946, he had to start collecting all over again.

This is where I came in.

As Red's first graduate student, I spent the summer learning the ropes, accompanying him on collecting trips and checking our booty back at the lab—only one fairly strong female strain had survived the Cal Tech disaster. We needed to collect stronger females and try to find a good strong male strain for testing female *Hormone A*. For a thesis project, I would try all sorts of conditions to persuade the females to produce hormone in very large quantities—more than Red got at Cal Tech. We hoped to harvest enough for chemical identification. Alas, as we learned only later, that was a project before its time.

Despite this venture's lack of success, it had significant side effects.

All-day collecting trips took us to freshwater ponds across Illinois and into bordering states. Carrying knapsacks packed with little glass sampling bottles, labeling tape, paper pads, pencils, sandwiches, and pickles, we trekked remote woods to distant ponds where few humans ventured. On these occasions, Red and I shared many thoughts—some of a casual nature, others bordering on intimacy. We got to know one another in ways few students and professors ever experience. Besides the science, we talked of common interests: the love of Bach and trumpets, our favorite books, the fun of camping out and climbing mountains. Then we spoke of family. While munching sandwiches beside a pond one day, I remember asking, "What was it like to be brought up on a tobacco farm down South?" Red described a kind of primitiveness I'd never experienced.

Red in plaid

He spoke of it like this: "Miserable, mostly—hot and dusty in summer; cold and drafty in winter. We had no plumbing or central heating. We ate what we could grow; never went to a restaurant or bought food from a store. Tobacco was the cash crop, but there wasn't much cash. In fact we were dirt poor but didn't even know it then, since our neighbors were just about as poor."

As I admired how his gray plaid shirt matched his gray-blue eyes, he went on: "During growing season, we worked the fields from dawn to dusk every day except Sundays, when we had to go to church—twice: Methodist church in the morning and Moravian church in the afternoon. Sunday was the only time off from work. I hated suckering tobacco most, especially after all my brothers left for better things to do elsewhere. I suppose the reason I smoke so much is I'm just trying to get back at it!"

We compared notes. My take on farm life differed from his; I thought of it as recreation. I liked bicycling down to the family farm (then run by my oldest brother) and helping with chores: pitching hay, picking apples, walking the woods, riding the work horses, trampling silage. I felt delightfully proud when my brother pronounced me old enough to drive the manure spreader all by myself—my favorite job.

Unlike Red's family, my family had time for fun things together: swimming in the lake, hiking the Adirondacks, coming home on a Sunday eve after a long day's skiing and lying by the fireplace, munching apples and popcorn while listening to Fred Allen and Jack Benny on the radio.

"My family read a lot," I explained. "We had shelves full of books and encyclopedias, took the *New York Times* and *National Geographic*." Red countered saying, "We had no reading material in our home but the Bible and a weekly community news bulletin. My parents were essentially fundamentalist. They considered card playing and dancing instruments of the Devil. Just the same, my father thought we all ought to grow up, get an education, leave the farm, and do something else for a living. I think he didn't like farming much more than I did. But the Moravians (on my mother's side) did like music; that's how I got to play the trumpet. In fact when I went to college I thought about taking up trumpeting professionally until I realized I'd rather be a second-rate scientist than a second-rate trumpeter."

We joked about the teasing each of us came up against as the youngest of comparably large families. Only he, one of seven boys, resented it more than I, the only girl. We learned a lot about each other's family during these collecting trips. One thing we shared was a strong sense of family loyalty and love.

I liked getting to know my mentor—his thoughts outside of science, his special kind of humor. I liked the way we laughed together, how he respected my thinking, admired my spunk. Oddly, Red seemed little older than I—a friendship was developing.

BEYOND THE COLLECTING, LAB work plus coursework left little room for anything else that summer. By spring, Ole and I had moved from a single rented room to a place of our own—an appalling basement "apartment." Its entrance led from a dark alleyway into a working cellar, complete with coal bin and coal-fed furnace. A rickety dining table perched at the back of the basement in what was meant to be a sort of kitchen. A small room to the north contained a toilet and overhead shower that drained directly through the basement floor. An enclosed space with two cots and two small desks lay to the east. It served as a combination bedroom-study, where the walking feet of passersby could be heard and seen as we looked up from our studies. We furnished our home with junk from Cottage Grove flea markets, acquired a little black pup from the pound, and delighted in the independence this glum home afforded. When Mom and Pop came for my

Graduation, University of Chicago

graduation, they did not admonish me or openly complain. It was only later that I learned how much this setting for two young girls in the big bad city of Chicago disturbed their sleep at night.

MY BROTHER LUTHER CAME back from the war in Europe that summer and started his doctoral work in International Relations at the University of Chicago. His expenses were subsidized by the G.I. Bill of Rights, a marvel of government wisdom that put many veterans of World War II through college and beyond. Lute, until then, had not been my closest brother, but as fellow grad students we began to bond. He grew up *being* an intellectual; I was only now trying to *become* one.

He, Ole, and I enjoyed rare free time together—swimming in Lake Michigan, bicycling to the zoo, listening to music.

I especially appreciated Lute's effective defense of my reluctance to embrace the communist ideology then peddled by my friend Lou. I had read the doctrines of Engels and Marx and knew their attractive utopian concepts: every human

Brother Lute with me

equally valued; each contributes according to ability; all share according to need—sounds like the teachings of Jesus Christ to me. But reality defies perfection, I thought—humans just could not live like that for long. Besides, as of then, it would start with dictatorship. Lute agreed.

Armed with knowledge beyond my ken, he and Lou debated point for point the history of communism versus democracy and the benefits of authoritarianism versus laissez-faire. Each matched the other on facts and dates; I listened and marveled while Lute articulated my gut feeling: Given human fallibility, government is only as good as the people who run it; unchecked power, no matter how benevolent at conception, poses danger to the governed. Despite its cumbersome aspects, democracy, with its built-in system of checks and balances and the possibility of periodic change, is perhaps the best we can do.

Lute bested Lou. In a way, he rescued me. Any notion I might have had to leave biology and concentrate on the political or social sciences faded entirely. I don't recall having any serious political discussions with Lou after that confab. In fact, to my relief, Lou gave up on me and soon found another girl. In retrospect, I wonder if he was the party mentioned in Harvey Tomlin's letter who "messed around" with Red's *Achlya* cultures for " . . . nearly five months and then went over to chemistry . . ." After all, Lou was a chemist at Cal Tech; he switched to political science when he came to Chicago after the war.

Lute became a pal and part of our gang who regularly dined at a little hole-in-the-wall down on Harper Avenue called Valois. Run by a likable family of French-speaking émigrés from Quebec, Valois welcomed a variety of customers from many walks of life: taxi drivers, cops, students, shopkeepers—all happy to take advantage of a limited menu of excellent food at bargain prices: homemade biscuits, chicken and dumplings, beef prime rib, everything good except the five-cent coffee. We liked this friendly mix of folks away from the University.

Lute also introduced us to the Hazlets, the only students we knew who lived in a decent apartment with good hi-fi equipment. Eight or ten of us shared our collections of long-playing 78s there at Saturday night parties. We each contributed a recording of choice—Mozart, Beethoven, Brahms, or the like—and listened all night to favorites played loudly while sipping cheap wine by candlelight—a scene foretelling romance.

Chapter Eight

Mad Love

ACTUAL EXPERIMENTS FOR MY master's project had gotten underway by that first autumn in Red's lab. We chose the remaining, strong female strain of *Achlya* in the collection, number 355, and started culturing her in ten-liter glass carboys fitted out with cotton-stoppered bent-glass tubing for aeration. Red taught me how to fashion the tubing using his rather simple glassblowing equipment: a blowtorch hooked up to a controllable gas source and oxygen tank. The air jet to which my intake tube was connected by rubber tubing was opened just enough to send a gentle stream of bubbles from bottom to top of the ten-liter carboy and provide constant circulation of 355's fuzzballs.

Some of those first few tries met with frustration due to interference from uninvited critters of the bacterial type, despite cautious application of sterile technique. It was one of my first lessons in the costly effects of carelessness, however minute, in doing experiments involving the pure culturing of biological specimens. In a letter to my mom I wrote, "My work is suffering a few upsets from bacterial contamination. I sure wish they'd leave you alone when you're trying to work with fungi!"

I CAN REMEMBER ONLY two courses during that first part of grad school: Sewall Wright's Advanced Genetics and Red Raper's Seminar in Mycology. Both courses influenced my future life as scientist in ways I could not then imagine.

Wright's lackluster delivery was countered by an honest enthusiasm that emanates only from hands-on experience. An established giant in the field of genetics, he was the first to work out a unified picture of evolution based on Mendel's principles of heredity.

Wright, a short robust gentleman with glasses and gray hair (perhaps resembling Mendel himself), aspired in his youth to be a professional baseball player but reached only the minor leagues. Now an esteemed geneticist, he was known not only for his innovative thinking about evolution and population genetics but for discovering relationships among a group of genes that determine the mottled coat coloring of guinea pigs. He brought guinea pig pelts and sometimes live specimens to class, to illustrate which genes in what combination were responsible for one patch of color versus another.

One time, he absentmindedly picked up a pelt instead of the eraser to make room for yet more of his numerous genetic symbols and mathematical formulae consuming the blackboard. I watched the show from the first row, having recently gained the courage to sit up front with the guys. At the end of class, he'd be covered in chalk dust, often stooping to retrieve those frequently shuffled note cards that somehow fluttered to the floor during lecture. Some thought Wright disorganized; I think he planned things carefully but would get so excited he'd forget the plan and digress into some relevant concept that happened to come to mind.

Always willing to entertain questions, Sewall had a hard time recovering the flow from such interruptions. Some answers were memorable. I asked one day, "How do you distinguish between wild type and mutant when the genetic material is changing and evolving all the time?" His answer went something like this:

> There's no clear line between the two. Think of a Swiss Army knife. It comes with a big blade, a little blade, corkscrew, leather punch. With use over time, the big blade gets nicked and perhaps filed down or replaced—likewise, the small blade. Then the corkscrew breaks and might be replaced with a screwdriver, and so on. Over the years, the knife's working parts change. In the end it looks pretty much the way it looked in the beginning, yet it has gone through a series of gradual transitions only some of which are easily detected. The exchange of screwdriver for corkscrew, for instance, might be recognized as a mutant form of that model knife, but otherwise the repaired present version resembles the wild type from which it came.
>
> In genetic terminology the array of genes specifying the character of an organism is the genotype, and the detectable character of an organism is the phenotype. We can think of wild type as the phenotype most frequently seen in nature. The wild type version changes through mutation of its genetic material, but the effects of those mutations may not be readily seen as change in phenotype; the organism is therefore not recognized as mutant.

In my future life as scientist, I often thought of Wright's answer to that question, especially now when much more is known about the entire genomes of many organisms. Genetic comparisons of individuals within species reveal a multitude of minor differences even among those who appear to be very much alike. We humans, especially, are all mutants in some degree.

Sewall Wright taught me to *love* the logic of genetics. Red Raper conveyed a *passion* for fungi.

As a less seasoned teacher than Wright, Raper lectured then in machine-gun mode. Mishearing a sentence or two would put the listener hopelessly behind. I had the unfortunate habit of arriving a few minutes late for class, whereupon I would try desperately to catch up by glancing at a fellow student's notes. One morning, early in the course, Professor Raper greeted this disruptive pattern of mine by stopping his lecture and addressing me formally: "Miss Allen, can you not manage to wake up five minutes earlier and get here on time?" The embarrassment effected reform—for that course at least.

The fungi, as taught by Red Raper, captivated my interest. They encompass a bizarre group of organisms filling an amazing diversity of niches in nature, from polar regions to warm pools, oceans to deserts. Even more surprising is the variety of ways the fungi propagate. Their multifarious lifestyles suggest that nature used the fungi as testing ground for all the different ways of accomplishing sex. Red Raper found that aspect particularly fascinating. So did I.

Red taught more about how fungi function than how they are classified according to looks. We studied a variety of the lowly creatures in lab, and in various experimental setups—*Achlya,* of course, but other water loving molds as well, bearing melodious Latin names such as *Saprolegnia, Allomyces, Dictyuchus,* and *Pythium.* We learned how to catch these critters in their natural habitat by sampling water from nearby ponds and baiting the samples with bits of nutrient the fungus preferred, such as morsels of boiled hempseed or autoclaved housefly.

Back in the lab, we purified our cultures, looked at them under the microscope, watched them move and change. This was my first look at differentiation of the vegetative spore-bearing structure of *Achlya,* and *Saprolegnia,* for example. The tip of a tubular hypha gradually swells, becomes granular with little round bodies inside, then suddenly pops at the top, releasing a mass of tiny spore cells, each of which sprouts a pair of hairlike flagella that whip back and forth to propel the spore away from its fellows in search of some substrate to support development into a fuzzball of hyphae. This whole pattern can be observed in a single sitting at the microscope.

I was thrilled to share a view of such odd behavior in the underwater world. The swimming spores look like little animals; the hyphal colony, more like a plant. Although these water molds are not now classified as true fungi, I still like to think of them as such, or to some extent related.

I'd never seen a slime mold before. Some slime molds come in brilliant colors, creeping like shimmering blotches over rotted logs or a cache of decayed apples. Under the microscope, an individual appears as one giant cell, full of big round nuclei that move together in streams, as with a tide. It is possible to make them move by teasing the cellular edge: poke one side with a pair of tweezers, and the nuclei flow toward the opposite edge. When the nuclei get ready to divide, they all do it at the same time. Upon some not-well-understood cue, hundreds, thousands of them divide all at

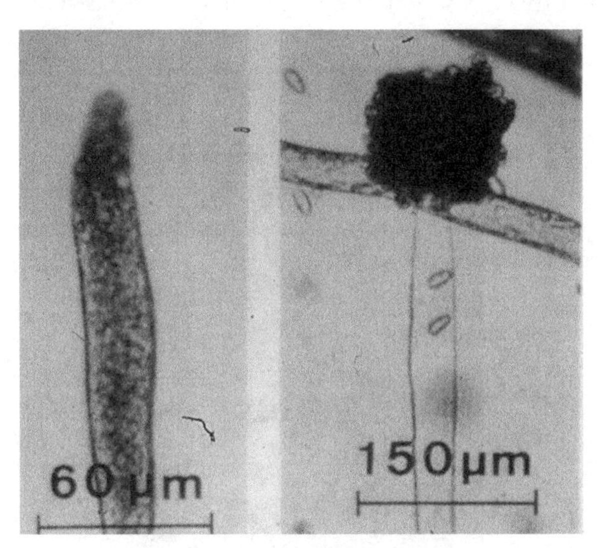

Achlya *reproducing without sex*

once. I wondered if this might be a good model organism for studying nuclear movement and division. I thought it would be grand to define the genes involved. Might I someday be able to do this?

Another of these strange life forms we examined is a fungus called *Pilobolus,* the "hat thrower," for which a modern dance company is named. We collected it from horse dung and watched it make a black hat of spores on top of a bulbous stalk (a sporangium), then, like a gun, shoot its hat toward a focused beam of light.

Bob Page, a former student of Red's at the University of Indiana, measured the velocity and accuracy of such shootings. He contrived an elaborate apparatus in which a flash of light triggered this fungus to snap its own picture as it shot its wad across a microscopic field. Results of several trials revealed amazingly high velocities averaging 35 feet per second. That's all the way from the kitchen through the dining area and living room of my house in one second's time! Not faster than Superman or even the speed of sound but faster than a dancer can leap across the stage and fast enough to

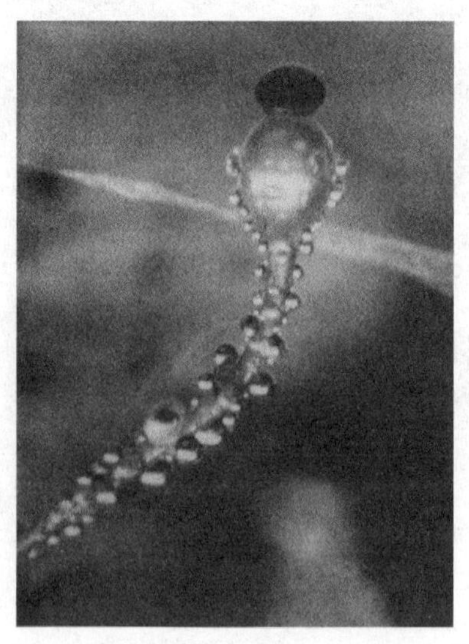

Pilobolus

spread a hat-full of spores through cracks in a covering of leaves or grass.

I admired Bob's ingenuity in designing such experiments. He parsimoniously assembled his picture-taking apparatus with castoff bits of equipment scrounged from other labs, even street-side waste barrels. Having quantified the speed with which *Pilobolus* throws its hat, Bob went on to study the forces causing it.

I later had the privilege of knowing Bob and enjoying firsthand his innovative thinking, his droll sense of humor. Always the *investigator*, he wore a magnifying glass strung round his neck in readiness to examine minute forms of life not easily seen with the naked eye. Always the *teacher*, he carried an enormous chunk of chalk stuck in his tweed jacket's breast pocket in readiness to illustrate a point at the nearest blackboard. Bob's perception of absurdity in himself as well as others kindled a succession of chuckles no matter what the subject of discourse. I loved talking shop with him; ours was always an exchange of listening and responding respectfully to each other's ideas.

Tragically, Bob died one night at the age of 49. His heart gave out while he soaked in a tub after one of his daily jogs around the Stanford campus where he taught. The shock borne by his survivors was tempered somewhat by the thought that Bob himself must have succumbed wearing a sardonic smile as he sensed the irony of dying after such health-giving exercise.

Bob Page was one of several whose skills and enthusiasm for discovering the weird ways of fungal life hooked me for life in a love for fungi. It was his widow Ginny who knowingly consoled me upon Red's death.

THE WEATHER, ESPECIALLY BEAUTIFUL that fall, encouraged long weekend bike rides—west to Garfield Park and the Brookfield Zoo and east, by train, for romps in the Indiana Dunes. I was happy with my friends; Red was not so happy—

lonely, in fact. On beginning the fall quarter, he had moved alone into a gloomy old rooming hotel somewhere over on 63rd Street. An acute housing shortage just after the war put promised living accommodations for the Raper family on hold for at least that fall. While their new home was being prefabricated, Ruth and newly adopted eight-month-old Billy went back to stay with her parents in North Carolina.

This forced separation exacerbated a problem Red and Ruth had been experiencing for a couple of years or more. It started in Oak Ridge where Red, working long hours on new projects under top clearance from the FBI, could not discuss his science with any unauthorized individual—including his wife. This ruling took its toll; the sharing of scientific thinking had always been an important part of Red and Ruth's marriage. Meanwhile, a darling, colicky baby boy joined the family, and naturally Ruth devoted most of her attention to him. Red and Ruth began to grow in different ways, develop different interests; the bond started to unravel.

Upon leaving the lab late one fall afternoon—October 3rd to be exact—I skipped down the stairs calling, "Goodnight, Dr. Raper!" He bent over the top railing and called me back. I thought I'd forgotten to do something I should have done before leaving, like shutting down the Bunsen burner or refrigerating his precious hormone. "You forget," he said, "I answer better to Red." Then, fixing his eyes on mine, "Do you have any special plans for dinner? It happens to be my birthday, and I don't know anyone here to share it with." I casually replied, "No, nothing beyond the usual meal at Valois with my roommate Ole and my brother Luther. Would you like to join us?" "Yes!" he said with apparent relief—a seemingly innocent exchange that marked the beginning of sorely troubled times for both of us.

Red enjoyed our company, and we, his. Among other common interests was a passionate love of music. Naturally he joined our Saturday night wine and music gatherings at Don and Connie Hazlet's. We also shared a love of sailing. Don and Con occasionally invited us to sail with them on their 26-foot ketch out of Burnham Harbor. The only one of our gang with four wheels, Red enhanced our common fun by jamming more of us than fit into his beat-up old Plymouth coupe for outings to the Dunes.

This all seemed harmless fun until one Saturday evening, while we lounged by candlelight, listening to the achingly beautiful Sanctus of Berlioz' Requiem, Red reached for my hand and squeezed it. I melted. A feeling of helplessness engulfed me. Had this been building from the moment we met? Getting to know each other through all those collecting trips out in the wilds? I did not dare think it—this married man, this father of a baby boy, my professorial mentor. It was not supposed to happen, not in that time, not in my family nor in his.

BY THE END OF fall quarter, when Ruth and Billy returned from North Carolina to join Red in their newly finished prefab, we two had fallen hopelessly in love, a form of obsession, if not insanity, totally unfamiliar to me.

Distress overwhelmed me. I wanted, needed, to get away. What to do?

A Christmas vacation back home in Plattsburgh provided some distance for introspection. The ethical choice was to quit graduate school and move far away. *Where would I go? Should I leave science? Find a job? Get into another graduate program at some other University?*

It was torment for Red and me to see each other daily and not share intimate thoughts. Despite the immorality of our feelings, neither of us had the guts to part. As winter progressed, Red made agonizing efforts to play the loving husband while honoring his commitment as Billy's loving father. But as one who habitually expressed true feelings in looks, if not in words, he found such efforts impossibly transparent. Few could appear so profoundly sad when sorely troubled!

Confessions to Ruth, who naturally chastised him, and sessions with an ineffective psychiatrist, offered no solution.

Divorce and remarriage seemed all too shocking—in fact taboo. There had been no history of divorce in either of our families. Red recalled a scandal so seriously troubling to his father Frank that he left the local church. The church's minister refused to oust Frank's sister who was rooming in the home of a man and his wife. While Frank eventually did rejoin the church, he never did forgive his sister, even though she married that man after his wife died.

Red's father, a Bible-loving fundamentalist, lived by church teachings. Red, as an adult, rejected the dogma but retained a large store of moral rigor. I was beginning to realize what a profound influence Frank Raper had upon his youngest son and how that legacy would lead to a period of intense anguish.

Chapter Nine

Water Mold Versus Mushroom Sex

MY THREAT TO LEAVE Chicago met strong resistance. Red countered with a plan. He would leave on a justifiable collecting trip to England and remove himself from both attachments. A few months away might allow him to choose between what he saw then as a passionless but morally correct future with Ruth, or marriage to me. In any case, he needed more specimens of *Achlya*, especially strong males, seemingly scarce in America but rumored to exist in English lakes and ponds. Perhaps the trip was meant as a mission to find two kinds of effectual males: one for either Ruth or me and one for *Achlya 355*.

The collecting trip appeared eminently sensible to Professor Krause, the Chair of Botany, and was thus approved along with necessary funding. Meanwhile, having learned the applicable techniques, I would finish my research project in Red's absence, trying out various media ingredients and culture conditions to enhance production of *Hormone A* by *Achlya 355* as she waited for her mate.

Of course, several months intervened before this plan took effect. It was misery alleviated somewhat by the advent of other graduate students in the lab and the company of an affable young man by the name of Bob Nunn, whom I met one evening at a party thrown by my roommate Ole and her boyfriend Jack.

Bob, vibrant, robust, and handsome, had fair hair cut in the current fashion to resemble a brush. We shared common interests as graduate students, he in an MD/PhD program in Physiology, I as a master's student in Botany. We talked easily from the start about whatever came into our young minds: He, for instance, found Aldous Huxley's satirical view of science and its effect on contemporary culture utterly fascinating. We read *Point Counterpoint* together and discussed it in depth. Huxley's opinions on science, politics, religion, and art were refreshingly different from those of Marx and Engels.

Unlike with Lou, Bob and I agreed more often than not. We respected each other's thinking. More comforting still, he, like a brother, patiently listened to tales of my agonizing love life. I wonder now if I wished then that Bob would take me away from it all. *Why couldn't I respond to* Bob's *pheromones the way I did to Red's?*

One particular graduate student in the lab provided another welcome distraction. Haig Papazian, fresh from the University of London, was to have a profound influence on Red's (and my) research far into the future. An eccentric fellow of Armenian extraction, Haig sported a shock of dark hair and a flamboyant mustache extending as upturned twirls on either side of his lip line, like some 1890s New York bartender.

Clothed in his usual jumper and heavy tweed jacket, he strode into the lab one August morning and stood before Red's desk. Red looked up quizzically,

"Well Haig, what do you want to work on?"

"*Schizophyllum commune,*" said Haig.

"The hell you say! Why that fungus? Hans Kniep worked out its sex life way back in the 1920s. What more can you contribute?"

"I've an idea there's more to it than Kniep worked out. Any organism with multiple sexes has got to have more than two genes running it."

Haig

"Well, you could be right—how will you start?"

"I've already started. I walked in the woods last weekend and picked up what looks like *Schizophyllum* on a fallen oak log."

Haig reached into his shirt pocket and fished out a small handful of tiny mushrooms. They looked like half scallop shells about the size of his little fingernail. Each had bifurcated ribs (split gills) on the underside—hence the name *Schizo* (split) *phylum* (leaf) from Greek. The *commune* nomenclature, I suppose, means "common" or "within a community." Haig, in his cultivated British accent, pronounced it, "skyzo-fill-um, ka-mew-nee."

"We can prove these are *Schizophyllum commune* by getting Kniep's specimens from the Dutch collection where he deposited them and check them out with mine—see if mine mate with his."

Red nodded, frowning, and gave doubtful blessing.

I marveled at Haig's verve and Red's willingness to accept

such declaration of independence from this beginning graduate student. Professors rarely grant their students such freedom of choice for a thesis project. More often, graduate students' research fits the professors' own preferences. This instance reaffirmed my admiration for Red as mentor, with an ability to subdue his position of authority and place trust in his student's judgment. That trust was not misplaced, for Haig's project launched decades of investigation into the genetic and molecular mysteries of mating and sex in the mushroom-bearing higher fungi.

Kniep was one of the first biologists to describe an organism with more than two sexes, actually many more. He got as far as figuring out that multiple sexes in the wood-rotting mushroom *Schizophyllum commune* have at least two genetic determinants, *A* and *B*, each with many different versions, and that a compatible mating required different versions of both *A* and *B*.

Haig extended Kniep's studies through numerous genetic crosses. He discovered that *A* is not just a singlet but has two parts, *A alpha* and *A beta*. He also showed that these linked *A* loci govern only part of *Schizo*'s sex life; *B*, also consisting of two linked loci called *B alpha* and *B beta*, does another part of the job.

Unlike *Achlya*, which, at the time, did not lend itself to genetic studies, here was a possible gold mine for getting at the genetics of mating in an organism easily analyzed.

Loving genetics, I was already frustrated by *Achlya's* reluctance to give up her fertilized eggs for development into adulthood—a necessary step for such analyses. Well, if Haig didn't carry on with *Schizo*, maybe Red and I would—some day.

(Above and right) Schizophyllum *on wood*

Schizophyllum *matings: incompatible (no mushrooms) versus compatible (mushrooms)*

Chapter Ten

Argonne Science and FBI Clearance

THE SUMMER OF 1947 arrived, and my master's project had hardly progressed. I'd learned the laboratory work ethic, but efforts to find a formula for increased production of *Achlya*'s female *Hormone A* produced only fractional success. Confidence in my ability to function as an independent thinker had diminished. Unlike Haig, I felt rather like a technician carrying out an assigned piece of research that was not going very well. My stipend would end in the fall. I would need to finish, then find a paying job.

To add to my woes, I'd perpetrated a grievous accident that could have shut down the *Achlya* project altogether: Red foolishly kept *all* the solution of *Hormone A*, laboriously concentrated during his postdoctoral year at Cal Tech, in a refrigerated 250 ml glass-stoppered bottle. Whenever needed for experiments, it would be transported carefully across that tiled floor and placed on the bench near our working space. One afternoon, when Red was out lecturing, I set that bottle at the edge of the bench right by the microscope. While I was examining a specimen, something diverted my attention—probably a cool bit of the hot jazz I often tuned into while working. Regrettably, my arm jerked in such a way as to make contact with the bottle. The bottle toppled to the floor and smashed to pieces, spilling its precious contents into a spreading puddle.

Red described the scene later for all to read as End Paper in a botany textbook called *Botany: an Ecological Approach.*

> During my first year at the University of Chicago, after the war years as a radiobiologist, work was continued on the hormonal action of *Hormone A* and the physiology of antheridial induction in *Achlya.* For this work there was available a pitifully small supply of *Hormone A* of high purity and standardized activity. This vial of standard was dear to my heart—it being used only in critical experiments and then only in 0.01 milliliter portions.
>
> Imagine my horror upon returning from a lecture to find my assistant, a fair-haired, first-year graduate student, on her hands and knees in the middle of the laboratory sucking up this precious liquid with a tiny pipette. She had dropped the bottle, which had smashed, and had intuitively gone about the rational business of recovering what she could of the hormonal solution with the equipment at hand. In a mixture of shock at the obvious carelessness on the one hand and my admiration of her initiative in making the best of a totally unnecessary and bad situation on the other, I could only urge the completion of the task and enjoin her not to cry over spilt hormone. There was, of course, no possibility of precise comparison of the activity of the recovered hormone with the original; it may well be that the quantitative aspects of the work with *Achlya* underwent a slight discontinuity as a result of the accident.
>
> Forgiveness, however, was apparently not too difficult. Perhaps my failure to erupt into the violent display of temper that had been suggested in earlier and far less serious situations convinced her that I might be human after all. In any event, a couple of years later we were married,

but over the years I've come to appreciate the moniker of 'Spilly' bestowed on her by her family at a very early age.

By the end of July, I had interviewed successfully for a job as Junior Scientist (meaning technician) at the Atomic Energy Commission's Argonne Laboratory, located in the poorly renovated horse stable of an old brewery, south across the Midway in a bad part of town. Although I had not yet achieved my master's degree, I was offered a choice of two labs, both working on the deleterious effects of mutagenic agents on microorganisms. Having opted for the protozoan *Paramecium* over the alga *Chlamydomonas,* I started as a paid employee and assistant to Dr. Larry Powers on the first of October 1947.

Meanwhile, Ole and I abandoned our ersatz apartment in the cellar. She moved in with her boyfriend Jack, and I found a room over on 61st Street, closer to my new place of work.

Red had boarded the *Queen Mary* and set sail for England one month earlier. The experimental work for my thesis was not yet complete. I would therefore have two jobs to keep me occupied while Red was away seeking virile male strains of *Achlya* as well as solutions to his own male dilemma.

Work at Argonne National Laboratory required the highest "Q" clearance by the Federal Bureau of Investigation, an astonishing process, particularly in view of the fact that nothing I would do or even hear about others doing, had any apparent relationship to matters threatening national security. Such precautions reflected a climate of fear, a switch from fear of fascism to fear of communism at the beginning of the Cold War.

Upon getting some feedback as to how my Q clearance investigation had been conducted, I wrote to Red:

> You should hear what the FBI man had to say to Anne and Jack about me. He was much impressed by my record, seemed terribly interested, thought I'm extremely intel-

> ligent (from interviews and records) but had one worry—that I knew Lou. Now how did he find that out? Who told him? I didn't. He also had some concern that my former landlords, the Wells, declared me a communist—those dirty rotten stinkers. I never ever opened my mouth on the subject of politics with either of them; but my books by Marx and Engels likely roused suspicion.
>
> Jack [an acquaintance of the Wells'] assured him it was a case of student curiosity, and the FBI man seemed to concur since he had gone through a similar period of interest in learning about communism himself. Furthermore, he was inclined to discount the Wells' statements on the grounds that he sensed they held some personal animosity toward me.

The male member of that couple had served as a super-patriotic Air Force officer during the war. He reminded me, in retrospect, of the kickass General Jack Ripper from the movie *Dr. Strangelove.*

The investigative process reached as far as my hometown of Plattsburgh, where my childhood friend's grandmother was grilled about my political integrity even though she had neither seen nor heard from me since high school. Most shocking of all was a question put to a co-worker the following year. An FBI agent asked how I planned to vote in the first presidential election for which I was eligible. Would I choose the Democrat "Give 'em Hell" Harry Truman; the ethereal socialist Henry Wallace; or the Republican guy who looked like a wedding cake doll, Thomas E. Dewey? I honestly didn't know whom I'd vote for until I reached the voting booth. My final choice helped carry Harry to an upset victory.

With adequate, but apparently incomplete clearance, I entered a work world quite different from that of student life. I wrote to Red about the first day,

> Got to work this morning at 8:30, and the gun clad, uniformed guard gruffly growled: "Where's your badge?" I had none. He said, "What d'ya mean you haven't got one? You can't get in here without a badge!" I dutifully trotted over to the University; waited around for a couple of hours to get it, and got back to work about 11:00 only to find the boss not there. His wife was having a baby, their fourth girl.

I later learned that Powers often left the lab, not just for domestic reasons but to serve as member of a Building Committee planning new and proper Argonne facilities west of Chicago along the Des Plaines River. But with two postdocs in the group, two other Junior Scientists, and a seasoned tech for making media and washing glassware, the lab ran quite well in the boss' absence.

The protozoan *Paramecium aurelia* put me in mind of *Achlya's* swimming spores but with larger dimensions and many more appendages. Shaped like a slipper with a concave

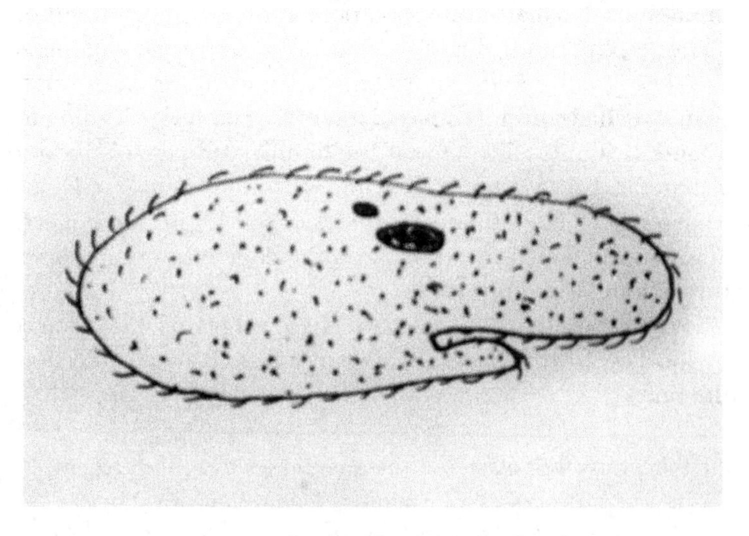

Paramecium *sketch with two kinds of nuclei*

groove serving as mouth, *Paramecium* moves through water like a multi-dimensional Viking ship. Rows of all-encompassing oar-like appendages called cilia propel it forward, sidewise, and backward.

Paramecium shares strange features with other related protozoa, such as two kinds of nuclei, multiple sexes, and the chameleon-like ability to convert its surface proteins from one serum type to another in response to environmental cues. Unlike *Achlya*, it can be manipulated for studies involving transmission of genetic traits. That feature alone attracted me. My major project was to compare the mutagenic and physiological effects of x-rays and nitrogen mustard on this rather cute but thankfully not cuddly animal.

But first Powers had me develop the technique of staining the beast in various stages of its development.

Paramecium aurelia is a single-celled animal found in freshwater ponds. Like all living creatures it makes more of itself by eating, growing, and dividing mitotically to make clones: one makes two, two make four, and so on in geometric progression. This method of reproduction, however, has its limitations: No matter how pampered, mitotic descendents of a single progenitor begin to slow down and wear out—like the cells of our bodies. Reprieve is possible, however, through sexual revitalization. *P. aurelia* evolved two pathways leading to a kind of immortality. One is the familiar strategy of finding a mate and recombining genomes through meiosis. (That's what we do.) The other is a process of self-fertilization called autogamy in which the animal, on the edge of mortality, goes through meiosis all by itself.

Powers wanted me to record these stages of mitotic and meiotic division by fixing and staining. As I wrote to Red at the time,

> I've pretty well mastered the techniques for two different intricate processes of staining *Paramecium* for Powers' collection. I hope he doesn't think my capabilities are indispensable to the point of wanting me to continue in

this vein indefinitely. I can see where it might very well become tedious after the novelty wears thin. I want to get on with the main experiments.

As for my master's thesis, only half of the first draft had been written by the time the job at Argonne commenced. I had to work weekends and evenings in Red's lab back at Barnes to complete the experiments with *Achlya* in his absence. I described one memorable day of disaster in a letter to Red:

Twelve hours work all shot to hell: I went over to Barnes to check on the bubblers as I had the day before and the day before that, expecting to be assured of their regularity. Five bubblers had ceased bubbling due to mycelial plugging at some unknown hour. Oh me! On top of that, contamination reared its ugly presence—Good Lord knows why! I was *so* careful! That leaves only two unaffected cultures out of twelve, and they look good. But what good are they when the comparison cultures failed? Everything went wrong, wrong, wrong!

I've been desperately trying this evening to make progress on thesis writing. It's going so slowly I'm beginning to suffer one of those *loss-in-confidence-of-self* moods, and with me, they're awful. To top it off, the weather's changed from hot to cold abruptly and my health likewise. I've developed a miserable cold, the raw coughy kind when you can't even smoke and food tastes awful. These complaints are genuine, damn it!

Oh, Sweet One, paper and ink are so very inadequate. I need your support. I'm confused and crave for complete understanding with you, about you, about us. Only frustrations with work keep me from thinking about you every minute.

Meanwhile, Red had his own frustrations in England.

Chapter Eleven

Painful Decision

HAVING JUST EMERGED FROM the brutal bombings of World War II, England had shortages of every kind in the fall of 1947. Red was shocked to find the populace shabbily dressed and food in short supply—an egg was luxury, meat extremely scarce, good cigarettes virtually unobtainable. He settled in at Cambridge University and shivered in a cold dank climate where heated buildings did not exist.

He wrote to me,

> Your first letter came this morning and knocked a big hole in some of the most complete aloneness I've known ever. I was dumbfounded that the place should look so poverty stricken—no laboratory glassware, limited supplies. . . .

That mood persisted. Two months collecting for *Achlya* in ponds and lakes around Cambridge yielded nothing. Red's sense of distress increased. He headed north toward the Lake District. Here the hunting seemed more profitable. As he shipped collections back to me, I checked each—using that floor-recovered hormone—for its degree of maleness. The samples looked encouraging.

Meanwhile Red had made progress with his personal muddle (he *thought*). In a carefully composed letter to Ruth, he wrote:

> . . . I have been away from Chicago now for almost two months; I welcomed this awayness as a chance to explore myself. . . . I still can't see clearly all of the facets to this thing and I despair of ever seeing all of them, but I can't help feeling that all along we, and particularly I, have felt too constrained to act, to think, and even to feel rationally. I at this time recognize the possibility of so consciously trying to force a rational program of behavior on rebellious emotions as almost certainly to preclude its accomplishment.
>
> As early as last Christmas you made the definite statement that you wanted either all of me or none of me—unless my attitude changed that we had best call it quits. Since that time we have repeatedly agreed that a life together without the old spark would be intolerable. Certainly I am not the me of old; there has been rather slowly a change in outlook and values and I want to emphasize most strongly that this change is not entirely the result of new associations and new relationships. In any event, the change was evident to me before external circumstances conspired to interfere.
>
> Where then does this morass of conflicting emotions and cross-purpose reactions lead? I suppose to this: separation for us. The subsequent readjustment, thereby made possible, provides for each of us the only real chance of any lasting happiness or emotional stability. I am not happy that I have finally come to the point of definitely suggesting divorce. But please let's have your unrestrained reaction to this.
>
> If in the recognition of continuing indecision and doubts, you feel as I do, that separation is the only way out, I would like for you to consider it along those lines and according to the schedule best calculated to mini-

> mize the inconvenience, discomfort, and embarrassment to yourself. Of course in the event that we do choose this line of action I shall insist on doing what I can financially and in every other way that will contribute to the comfort and security of yourself and Billy—Oh God, were it not so pitiably little!

Ruth had written earlier,

> I am now facing reality. You're gone and the guy I have fought with for the past six months is missed. I no longer think of us when we were happy for ten years. I think of us now. . . . My remorse is deep that not once have I shown any appreciation, any sympathy, and any love for the present you—only abuse and condemnation. Forgive my littleness and selfishness if you can. . . .

After receipt of Red's letter suggesting divorce, Ruth agreed to make the break immediately and do it as quietly as they had gotten married ten years earlier. She pleaded with Red to " . . . Please stop feeling sorry for all of us—and stop feeling guilty. If you can do that, I think you will get to be your old sunny self again and your troubles will resolve themselves into whimsies."

Such rationality did not prevail without interruption by continued feelings of guilt on Red's part, and bitterness on Ruth's. During the next several months, only two close friends and selected family members were informed about our tangled triangle of emotions.

Red went on with his collecting in the Lake District, sending me specimens for testing. Eureka! Two strong males emerged from the Lake District collection. That worry, at least, abated.

Yet our misery in absence of each other prevailed and intensified.

Now knowing of Red's decision about Ruth, I wrote:

I'm more lonesome for you than ever, especially at your lab—no lights, no whistle, no Red there to greet me with a twinkle when I arrive a *few* minutes late! I just tore away from [Maugham's] *Of Human Bondage* sleepy eyed, somewhat weary of reading about story people and their intense emotions. The damn novel makes me think too much about us, me, you, and life. My unrelenting love for you exists more in parts than in whole, and I can't seem to silence a question—will it, can it ever be realized as whole and permanently so? You've become a part of me; I'll never be able to forget it, and I really don't think I want to!

Whether or not our love is great enough to weather the social taboos, the difference in age, the feelings of guilt, I do not know now. I feel almost certain it is, but still, in all honesty, I can't be sure until we've experienced some test of time together. I do know that you are everything I want and need as companion and husband.

When will you be coming back *home*—such as it is?

Chapter Twelve

Frustrations

A MONTH REMAINED BEFORE Red's return, and I hadn't completed my thesis. I realized too late that the experimental procedure could and should have been more logical. Conclusive results seemed remote at best. In desperation I resumed a weeklong routine of evenings at Barnes in early November.

One night, near the end of a long experiment involving a series of ten-liter cultures that had been bubbling, I arrived at the lab to find everything eerily silent. All my cultures sat menacingly still. The building's forced-air supply had ceased without warning. Yet another setback—how to cope? Since this mishap occurred sometime during the tenth and eleventh day of an anticipated twelve-day run, I worked all night, dog-tired, trying to salvage the remains. I filtered each of a dozen cultures and tested the relative quantities of female hormone produced in each by using Red's new English male strains as responders.

Perhaps my ability to tough it out had its origins in childhood, when older brothers tested my mettle by ghoulishly swinging a fragile rope bridge over a rushing stream while I clung for dear life trying to cross from island to mainland with a bagful of butternuts. Or maybe it came from a face-

New little sister

full of mud when those same brothers pushed my head to the ground after I snapped the football to quarterback—it seemed I always played center. Whatever the game, they would nonetheless comment, "That's pretty good—for a girl!"

In truth, these taunts elicited laughter more often than whines. I knew they were trying to toughen me up. In fact, I almost lost hold of that swinging rope from giggling so much; they knew when to stop—just in time. Still, these provocations shaped and strengthened me—from birth—as only brothers could.

Results from those final foiled experiments proved erratic at best. None were publishable or particularly useful for hormone production in large amounts. They did, nonetheless, become part of my thesis as learning experience. I completed the document and mailed it to Red by the end of the following week.

With my thesis done I needed diversion. A friend persuaded me to attend a party thrown by a Communist cell. I must have wanted to live dangerously in confrontation with the FBI. As relayed to Red, it wasn't much fun:

The party was held in a cellar apartment, almost more cellary than my old place, with red walls, the hammer and sickle motif decorating the doorway, a prominently displayed collection of the usual leftist books and recorded music consisting only of Red Army Chorus songs, Union songs, songs for Democracy, and the like. There must have been about 75 people there of all shapes and sizes—buzz, buzz, buzz, wonder why they duzz! And *new looks*, my word, I contrasted strongly in conventional dress amidst an abundance of bare shoulders, shoulder-length earrings, Pepsi Cola hair-dos mixed in with equal numbers of work shirts hanging over dungarees. The folks were all quite new to me but cordial. It was interesting, spasmodically. It helped to drink beer.

RED SOUGHT DIVERSIONS OF his own. He wrote of meeting Brenda, a friend of a friend from the States. She and Red composed an empathetic pair. He needed female nurturing; she, of a nurturing nature, needed to be needed. They developed a mutually enjoyable friendship dining together, going for walks, attending concerts—all innocent enough at the time, only to have that friendship blossom into something more many years later.

Now toward the end of November, I too longed for companionship with the opposite sex. Graduate-student friend Bob Nunn and I liked similar things: classical music, jazz, strolling the lakefront, window-shopping down in the Loop, discussing imponderable questions in science. (By then, feeling uncomfortable in that place over on 61st Street because of the FBI incident, I had made my fifth Chicago move to a room on East 54th.) After a night of hot jazz at the "Blue Note"—Ella Fitzgerald and Satchmo in their prime—Bob and I got back to my place so late I dared not rouse my new landlady to let me in. (I'd forgotten my key.) We had no choice. I went home with Bob. He had only one bed. I teetered all night at the edge of his mattress, keeping space between us, longing for Red, trying to think of Bob as a brother while denying an urge. Bob nobly kept his distance.

I wondered later whether Red's encounters with Brenda and mine with Bob might have happened at nearly the same time.

Red returned from England in mid-December. Upon landing in New York he visited the Rockefeller Institute to request assistance in purifying and identifying *Achlya*'s *Hormone A*. The appeal met with no success. Ely Lilly also turned him down. As Red put it, "I guess they figured that identification of a female sex hormone in *Achlya* was of vital interest only to male *Achlya*."

The curious predicament of how and why strong males and females of the species got separated by three thousand miles of Atlantic Ocean remains a mystery to this day. Our failure, however, to identify the exact nature of *Hormone A* was taken up two decades later by a friend and colleague Alma Barkesdale, who, along with a collaborator using new techniques unavailable in the 1940s, discovered that both hormones *A* and *B* are steroids known before then to exist only in higher animals.

Red and I—starved for each other through long separation by that same Ocean—met impulsively for a weekend together in cold, snowy Cleveland. We arrived by train: he on Friday from New York, I on Saturday from Chicago. We had planned to do a number of things, but just being together again meant bliss. Wanting to share music, we hopped a snow-laden trolley to attend a performance of Bruckner's Eighth Symphony, a work neither of us knew before.

I remember not what was played before Bruckner, except that it was lovely and we held hands—after all the Cleveland Orchestra under George Szell was known for its excellence. But Bruckner's Eighth began to wear. It just went on and on without apparent purpose, disrupting our aura of warmth. Impatient with the tedium, Red and I exchanged looks of despair. If only Bruckner had had an editor! It was the one time in either of our lives when we abandoned paid seats and left a concert early for the warmth of a shared bed.

BACK IN CHICAGO, JUST before Christmas, Red settled into a dismal room within a dreary boarding house, its only attributes being cheapness, nearness to his son Billy, and proximity to the University. The landlady called Red *Yonnie* in a strong Polish accent and insisted that *Yonnie* keep his door open whenever he had visitors—no *hanky panky* in her domicile!

I continued with my work at Argonne, having been tapped by the Boss to type up his presentation for the upcoming meetings of the American Association for the Advancement of Science to be held that year just after Christmas at the Palmer House, downtown. I had by then managed to assemble a series of slides illustrating all stages in the life cycle of *Paramecium aurelia.* It was to be the central focus of Powers' talk at the meetings. I noticed my name was missing as one of the authors. This set me brooding: *Why should I, having done all the work and a good amount of troubleshooting in the process, not be acknowledged?*

After a restless night, I assembled the courage to confront Powers the very next morning in the privacy of his office. A moment of tension ensued. He paused. He blushed. He finally agreed to include me as an author. I don't know how Powers felt about it afterwards, but I felt good. He had treated us female techs with respect, but I noticed he listened somewhat more attentively to the males in our group—in all fairness, perhaps not *because* of their maleness but in view of their more advanced professional status. But then again, maybe gender did play a role, if inadvertently. Just the previous week Powers had hosted a famous visiting scientist, a male Nobel laureate, who "complimented" us girls in the lab by saying, "I didn't realize such competent young scientists could be so beautiful!"

In any case I had learned a lesson about accreditation in science: sure, Powers headed the lab and assigned this project in the first place. But I carried it out pretty much on my own and contributed a few ideas in the process. This was a first step in thinking of myself as not strictly a technician or just a pair of hands. I was beginning to feel like a real scientist capable of making independent contributions. That feeling would be fortified later while I worked on the main project for which I was hired.

CHRISTMAS CAME AND BROTHER Lute and I headed back home to Plattsburgh for a weeklong break. The train ride together seemed the right occasion for me to confide my strong love for Red. He, not in the least surprised, was clearly supportive, but agreed we should wait a while to tell Mom and Pop. Lute liked and respected Red. His empathy for my predicament vis-à-vis our parents may have been influenced by his own long-suffered dilemma of being gay within a society utterly intolerant of homosexuality. He dared not reveal his sexual preference to any member of the family other than me at that time. If it were to become common knowledge, his future would be sadly jeopardized. Damn such bigotry—much worse then than now. We both had secrets to keep.

Meanwhile, my love life remained unresolved.

Chapter Thirteen

Uncertainty

BACK IN CHICAGO AFTER Christmas, Red and I met with despair and deep uncertainty. His resolve to leave Ruth had seriously faltered with the tug of two-year-old Billy, who'd welcomed his Daddy back from England with endearing need. We agreed again to keep apart until Red could somehow come to terms with this new reality of conflicting emotions.

The agreement was broken within a week. I had just toweled off from a steaming hot shower and gotten dressed when, gazing out my bedroom window, I saw him there, out front on 54th Street, against the snowy sidewalk. Clad in brown with matching fedora, he looked up longingly. I waved, smiled; he waved and motioned a desire to come on up. I nodded consent. We were together again.

Red was on his way to deliver a talk at the Association of American Scientists down at the Palmer House, the same meeting in which my boss Powers would present my stained illustrations of the lifecycle of *Paramecium aurelia.* Red wanted me to join him. I did; I couldn't help it. We walked arm-in-arm to the Illinois Central station and sat in tandem on a train to the Loop. Just being near him quickened my pulse and made me feel happy.

Upon arrival we sat near the front of a crowded meeting room. I swelled with excitement as he walked to the speaker's podium. All went smoothly until one of his slides illustrating a graph of crucial data, some of which I had generated, appeared upside down. Trying to stay within time constraints by not pausing to have the slide reoriented, he proceeded nonplussed by making the point, "Even upside down one can see male *Achlya's* increasing response to increasing doses of the female's *Hormone A*." The audience chuckled at such aplomb. There are seven wrong ways and only one right way to place a slide in the projector for viewing—a problem all but forgotten with computer-generated slide shows.

After sticking around to hear Powers' talk, Red and I walked hand-in-hand to the warmth of a nearby bistro on Randolph Street. Brushing knees amid the aroma of our favorite lunch, steamed clams, fries, and beer, we knew all over again that nothing had changed; we needed each other as much as ever.

Still, Red's inability to deal with his dilemma persisted throughout the next seven months. I had fulfilled the requirements for the Master of Science degree and therefore had no working excuse to visit his lab. With some exceptions, we somehow managed to resist the temptation of meeting anywhere else.

It was during the summer of that year that I almost gave up altogether. *Will Red's instability persist indefinitely?* Our affair with its ups and downs seemed like an endless Mahler symphony. I felt like retreating. Even if he and Ruth should come to divorce, would our guilt be so strong as to preclude any true happiness afterward?

My despair hit bottom on the July 4th weekend when Ole, her Jack, Lute, and I vacationed at a friend's gorgeous summer place up north in Wisconsin. Fond memories of happy childhood celebrations with firecrackers, pinwheels, and rockets left me vulnerable. Dad would give each of us little kids a dollar to spend—the big boys more. We eagerly spent a whole day in advance, choosing the best for our bucks—Jonny preferred Cherry Bombs; I liked the less menacing Snakes and Lady Finger Crackers. Come evening, we'd drive to my Uncle Paul's farm up in the Adirondacks. We picnicked on the porch with dozens of relatives as the big boys shot off sparkling fireworks over by the pond. It was all joyous fun—except the one time when a rocket backfired in brother Fuller's hand and seared his skin to the elbow.

July 4th, 1948, however, despite its gorgeous weather, good food, and company, put me in a blue funk. The pleasures of the day, seemingly enjoyed by everyone else, enhanced my dark mood. My thoughts focused on a love affair I feared would end, as did the hopeless love between Laura and Alec in Noel Coward's *Brief Encounter*—never to be fulfilled, but to live only in memory.

MEANWHILE RED LABORED IN the lab, trying without success to persuade *Achlya's* fertilized eggs to germinate and generate, a necessary first step in finding genes for sexual potency. Nothing worked.

Red's student Haig Papazian, however, had made good progress with *Schizophyllum.* While struggling with thoughts of leaving Ruth for me, Red also considered leaving *Achlya* for *Schizophyllum.* Partly at my urging, but mainly through the influence of Sewall Wright, Red had come to realize the importance of genetics as a tool for understanding how organisms function. He began to work with Haig in collecting more specimens and figuring out the best way to analyze them. This was only the beginning of a long period of research in the Raper lab that started with Papazian in the late 1940s and persists in the labs of my protégés to the present day.

By the end of summer, Red and I got back together. Our turbulent love affair reached some kind of finale. Red and Ruth agreed to call it quits. Ruth filed on the grounds of desertion, but nearly a year passed before Red had healed enough to legalize a second union—ours.

Chapter Fourteen

Assertion

Lute and I went back to Plattsburgh in August for a three-week vacation with Mom and Pop at the family camp on Lake Champlain. It was at this time and in this setting that I informed our parents about my feeling for Red and his impending divorce. Dad frowned and Mom registered no surprise. I never could fool her. She, and perhaps more reluctantly Dad, took some comfort in Lute's enthusiastic endorsement of Red for me.

Red planned to stop by and meet my folks on his way back to Chicago from North Carolina, where he was visiting his closest family members to tell them of his intentions. No word had come as to his time of arrival. The camp being phoneless, I expected a letter or telegram. I worried as never before that something had gone wrong: a change of mind? an accident?

But then, while finishing a simple supper of salmon and peas on toast, we heard a knock on the porch door. Mom answered it, exclaiming "Who are *you*?" as if addressing someone looking for a handout. There stood Red, rumpled and travel worn. "I'm Red. Didn't you get my telegram? I sent it to the office." This marked the beginning of a bad relationship between my future husband and my brother

Fuller, who, as Dad's partner in law, presumably received the telegram but failed to pass it on.

"Why, Dr. Raper," my mother blurted, "come on in. We wondered when you'd arrive! Come sit down. Have something to eat." Once identified and embraced by me, Red joined the family fold and began to feel at home. Mom and Dad dared think of him as future son-in-law. His thank-you letter upon return to Chicago conveys a feeling of acceptance:

> Dear Mrs. Allen,
>
> Coming back to the beginning of a new quarter and to the initiation of a course that I had been allowed to forget for almost two years, hardly justifies my delay in writing you and thanking you and your family for the most restful and peaceful ten days I have had for years. Such it was; I enjoyed every moment of the time I was there. I can hardly say that Plattsburgh as a town lived up to the joint impressions of it that Cardy and Lute gained during their obviously happy childhood there. No place could be quite as romantic and fair as their away-from-Plattsburgh idea of Plattsburgh. I'm much more in sympathy with their deep affection for their family and I really feel that the locale of their early years could have been any other, including Indianapolis or Knoxville, with about the same deep attachment to it.
>
> I suspect we shall all be gaining weight. Cardy, Lute, and I have been eating dinners like kings since Cardy began running her exclusive dinner club. To watch her prepare a meal, her assertion that she can't cook is not too hard to reconcile to that state of familiarity with a kitchen, when even a pinch needs a definition, but the illusion of her being a beginner vanishes when it's time to eat!
>
> Again may I thank you for the very enjoyable visit I had with you and the many things you did to make it so.
>
> My very best wishes to all of you. Sincerely, Red

THE JOB AT ARGONNE began to get interesting later that fall when I started concentrating on my major project. Mine was part of a study on the comparative effects of two lethal substances relevant to warfare: ionizing radiation (as in atomic weapons but delivered here by x-irradiation) and nitrogen mustard (the ingredient of mustard gas used as a caustic weapon in World War I). Both substances have since been used as weapons against cancer. Because it is easily manipulated in experiments quantifying death by toxicity versus death by genetic mutation, *Paramecium* is an ideal organism for such studies.

My fellow Junior Scientist Daila Shefner worked on the x-irradiation part; I worked on nitrogen mustard. Daila's data showed a standard response: X-rays caused more and more deaths *and* mutations as she increased the dosage. My data delivered a surprise: The same damaging effects increased with dosage up to a certain point, then, unexpectedly declined. How can that be?

I couldn't understand how or why increasing doses of nitrogen mustard caused reduced amounts of both death and mutation at the highest doses. Having recently experienced the vagaries of *Achlya,* I doubted my results. But after repeating them in two subsequent experiments, I had to believe.

It was time to troubleshoot; *What could I be doing wrong?* Our boss Powers was away at some conference or other, unavailable for consultation. I had to *think* on my own. Ah ha! Would it make a difference if I changed the method of administering dosage? Instead of making up one solution and exposing the beasties for different periods of time, what if I did it the other way around? Make up fresh solutions of increasing strength of nitrogen mustard and expose each group for a standard length of time?

Sure enough, my second method of increasing dosage produced a response that continued upward and tapered off without peaking. *What could this mean?* Maybe the nitrogen mustard was changing from the time it was first dissolved.

A solution to this mystery required some joint thinking

involving Powers and a biochemist by the name of Pomeroy in a lab upstairs. Pomeroy knew that nitrogen mustard, once in solution, goes through a series of modifications. Apparently the first hydrolysis product packs the punch, while the second has virtually no effect at all.

The strange peak and fall of deleterious effects in my earlier experiments could be explained by degradation of the potent first product and its replacement by a developing less potent—or possibly healing—product over time. So the period of time that nitrogen mustard stayed in solution explained the difference in results. Interestingly, this response pattern seemed to be unique to *Paramecium.* It had not been seen in comparable studies with other microorganisms. This was new information, possibly valuable as a basis for future investigations.

In any case, it led to my very *first* publication: I was named third and last author. While last meant least in those times, I nonetheless felt pleased that this time I did not have to plead for authorship.

This challenging experience did wonders to bolster my confidence as budding scientist. I dared think that someday I could plan and interpret my own experiments without continuous direction from some higher authority.

It took another nine long months, however, before I'd become Mrs. Raper.

Chapter Fifteen

Marriage

WELL, WHEN IS THE wedding?" asked my parents when they came through Chicago on their way back from a trip west in May 1949. "I packed an appropriate dress and we brought you a Navajo rug for a present," said Mom. They seemed disappointed when we shrugged our shoulders.

Red by then had found a pleasanter place to live, albeit small, over at 54th and Woodlawn. We prepared dinner for the folks that night in our Pullman kitchen and ate in the so-called dinette, just big enough for four to squeeze around a second-hand table—extant today in our daughter's kitchen. The exquisite taste of grilled salmon and steamed artichoke leaves drenched in melted garlic butter surprised Dad, who had never known me to cook before. We had won them over. Mom and Pop approved of our partnership and patiently awaited its legal resolution.

PROMOTIONS AT WORK AT the end of June gave us the confidence to set a wedding date. Red became Associate Professor with a raise from $5000 to $5500 annually and I received a merit raise from $3000 to $3300 at Argonne. Happy times! I wrote my folks a 14-page letter from Chicago explaining our plans: " . . . Would Tuesday, August 9 suit you? . . . We think it would be nice to have the wedding at camp on the lawn near the shoot of the rose bush under which you were married."

I then presumed to specify numerous details of just how the ceremony should be: simple and informal, with no rehearsal required; mid-afternoon with only immediate family in attendance (no more than three dozen Allens and Rapers); larger reception afterwards restricted to an additional 14, making no more than 50 total; afternoon dress only; sandwiches and cake; and so on.

Well of course such explicit instructions shocked Mom. This was to be the only wedding she'd get to host and she wasn't about to have me call all the shots.

We fired a volley of letters back and forth—Mom, rejecting the ridiculous notion that the wedding ceremony and reception be separate with different lists of invitees. With Lute's counseling, I finally saw the light: This was Mom's party, not mine.

We happily married under a basswood tree by the lake on the hottest day of the year. The ceremony drew to dramatic close with a punctuating clap of thunder just as the family minister pronounced us *man and wife.* I since wondered, why not *husband and wife*?

RITUALS COMPLETED, RED AND I enjoyed a couple years of independent married bliss before seriously contemplating children of our own. Previous fears of guilt casting gloom upon our union had dissipated. We enjoyed Billy's visits; he seemed to be adjusting without trauma, but we knew it was hard for him not to have a Dad's constant presence. Thankfully, Billy accepted and liked me. He shared our pleasure in training a newly acquired wirehaired fox terrier puppy named Soo Pooh, as well as the fun of sailing in an old wooden sailboat we had bought halfway.

Life for me seemed better than ever.

Nell and Ben visiting us in Chicago

My mom and pop with the finally marrieds

Billy and Soo Pooh

Chapter Sixteen

Chicago to Harvard

THE BIRTH OF OUR first child in 1952 interrupted my position as Junior Scientist at Argonne Labs in Chicago. Like most women of the time I never questioned the concept of being a mother once married. I only worried some that I might not be able to conceive and have a normal child.

Seven pounds, six ounces, a healthy baby boy arrived conveniently on Father's Day in 1952. As he lay naked on my tummy, a perfect tiny thing, his head full of crazy hair, I wept shamelessly, melting with motherly love. We named him Jonathan Arthur after two fond brothers. The day before his birth we had sailed on Lake Michigan, packing a jackknife along with lunch, in case the need should arise to sever the umbilical cord. No need: Jonathan, the little sailor, waited until morning.

I now had the baby I wanted, fast and easy—except for the tyrannical obstetrician's insistence on a spinal just before the last few pushes. Doctors were then king. Mine not only permitted no choice of anesthesia but also issued an experimental medicine without explanation. Reputed to be the best in the trade, he insisted I take some large red pills upon threat of dismissal from his prestigious free clinic, should I refuse. "What are they for?" I asked. "To test their efficacy in preventing miscarriage," he answered. "But why me? I've never experienced miscarriage!" "Nonetheless, you're part of the test," came his reply.

How could prospective parents with scientific training, like Red and me, have agreed to such a travesty? I can only say in self-defense that this doctor was as intimidating as an army general, and we weren't wealthy enough to pass up a no-cost deal. In retrospect I still feel guilt, even though our son suffered no ill effects. Perhaps I was part of the sample that got a placebo instead of the test-drug diethylstilbesterol, although I was refused this information even 20 years later.

Results of this experiment, conducted similarly in major hospitals throughout the country, later showed that diethylstilbesterol did not prevent miscarriage but did cause an increase in cervical and vaginal cancer among mature female offspring born of mothers given the drug. Stringent constraints on the science of health care were sorely missing in the 1950s.

HIGH HEAT PERSISTED FOR most of the summer after Jonathan's birth. As a homebound mother I was granted a successive six weeks at half-time salary to take care of my newborn and write up my research for publication, altogether a generous benefit compared to many other institutions at the time. Paid leave for longer periods of time would not be common until the feminist movement was well on its way some three decades later.

Despite the gladness of having a robust son, I experienced a bewildering depression. It seemed impossible to accomplish anything—feed the baby, change the diapers, do the dishes, feed the baby, make the bed, baby again, make lunch for Red and me, nap time (?), more dishes, more baby. House-mom duties never ended; they just circulated. Lusty cries pierced my sleep. *Why can't Jonny be content?*

Finally a relatively new invention, a rubber nipple faced with plastic called a pacifier, came to the rescue. It seemed our baby had a stronger sucking instinct than could be satisfied by nursing. Might this reflect his parents' miserable habit of sucking on cigarettes? My mother disapproved of both cigarettes and pacifier—another matter about which to feel guilt, in retrospect. In any case, this Jonny demanded attention. Fortunately, Soozy, the terrier, greeted our new addition amicably and, as pre-agreed, Red learned along with me the diaper-changing routine—a rare practice among new dads at the time. It pleased me to learn some 30 years later that this very son changed his own sons' diapers—a not so rare practice these days as formerly.

Still, my depression persisted. I cried. *Why? Why? We had the wanted son, good marriage, financial security.* This shift from nine-to-five scientist to day-night parenting, on top of the writing, had unexpected consequences: Lab work, while sometimes frustrating, seldom bashed my spirits so much as full-time homemaking with baby on board.

I learned only later of commonly experienced depressive syndromes known as "baby blues" or postpartum depression due to precipitous changes in hormonal balance after birthing. Could that have been the cause of such discontent? No one had warned me. Did those old-time doctors think that some women just went crazy facing motherhood and nothing could be done about it? Alas, the answer is yes. Experimentalists had not yet conceived of a search for physiological causes. That would not happen until the 1980s, when women became more assertive about solving female-specific health issues.

NOW AS FAMILY, WE thought despairingly of raising children in Chicago. Affordable row houses near the University imitated one another in their contiguous ugliness and tiny back yards. Chicago's public schools had so little appeal that most of our colleagues with school-aged children paid the cost of private education. Although the University of Chicago's lab school had a fine reputation, we were conditioned to favor public schools where kids could mix with others from different backgrounds. Yet Chicago politics posed a seemingly insurmountable barrier to public school improvement.

Our outlook improved by midsummer, when a feeler from Harvard came in the person of Cap Weston, Red's cherished doctoral mentor, who was soon to retire as Professor of Botany. According to customs of the day, a retiree could actively recruit his own successor. The "old boys' network" prevailed, in which known protégés were preferentially sought. Now, strict rules for equal opportunity apply, requiring published ads in professional journals—any qualified person may respond regardless of race, creed, color, gender, or who knows whom. A network still operates nonetheless—known prospective recruits with desirable qualifications get heads up.

Of course I'd heard much about Cap from Red; yet when he arrived at our door for dinner I was ill prepared for his utterly enchanting nature. This tall handsome redhead (now grey) bowed and addressed me as Countess Carlaina. I was not immune to his flattery. He loved martinis as well as we. The cocktail hour extended beyond intention while ol' Cappy, in thespian mode, entranced us with gossip and tales of amusement. Meanwhile, the steak overcooked. No one minded; it tasted just fine in our jolly state.

The prospects Cap brought lifted me out of the doldrums, although his visit may have coincided with a new, healthier hormonal balance.

A phone call came late in August. Could Red come to Harvard and meet potential colleagues? Well, *yes*! We drove to Plattsburgh for a short respite at the Allen Camp on Lake Champlain. There we invited Mom and Pop to accompany us to Boston, followed by a five-day trip up the Maine coast. The weather cooperated to make this a glorious vacation. We all stayed at Boston's venerable Parker House, two-month-old baby, dog included.

My folks cared for Jonny and the dog while Red met with potential colleagues and Cap ushered me on a tour of the campus. Cap and I joined Red and two or three other biologists for lunch at the Faculty Club. I was blissfully unaware that my presence as female required use of the family dining room accessed through back door only—the front entrance to the main dining room being reserved exclusively for males. It would be another decade before that rule changed in response to feminists' protestations.

I cannot recall Red's account of his meetings with various members of the Department that day. It must have gone well, since he later was called to join them permanently.

Harvard, in its wisdom, required a year to come up with an offer: a tenured position as Professor of Botany, at $8000 per annum plus benefits. The appointment would start in January 1954. We learned only later the reasons behind so long a process. One, revealed by a third party many years hence, was that Red's brother Ken Raper received the offer first. After due deliberation, he declined in favor of a position at the University of Wisconsin, Madison. Ken modestly never mentioned this to Red. In any case, he always seemed content at Wisconsin, where he fared financially better; state-supported universities were beginning to attract good faculty by offering higher salaries than private institutions at the time.

A second reason for delay was the elaborate process by which Harvard approves appointment to tenure. First, the Department seeking such appointment must agree on the candidate, then the University Senate must agree to the suitability of the candidate's credentials, followed by approval of the Provost. But that's not all: An ad hoc board of professors in related fields of expertise must be assembled from all over the world to judge whether or not the candidate's subject of endeavor is important in the greater scheme of things and whether or not the candidate is the best obtainable in that cutting-edge field. Should one make it through this labyrinth, he must gain the President's seal of approval.

When Red later understood this convoluted process, he

thought it uncanny that *he* had survived to become a Hahvahd Professor.

WE THREW OUR OWN going-away party. Others entertained us with movable feasts and more formal gatherings, but we decided to say goodbye to Chicago friends and acquaintances as hosts in our own humble place. It being January, the windows stayed shut and festive candles plus all those bodies in three small rooms sucked up the air. Toward the end of the evening, I revealed my condition as three months pregnant by dramatically fainting in the clothes closet as I reached to retrieve the Departmental Chairman's winter coat.

No worry. We were on the road to Boston within the week.

WHILE WE SETTLED INTO a rented apartment, it took six months to find our dream home, constrained as we were by a rather meager budget. Anything suitable near the Harvard campus proved far too expensive. After numerous rejections of suburban cape and ranch style houses, the realtor in her wisdom showed us an architect's replica of an old Cape Cod style house with a bowed roof resembling the hull of a boat. Painted dark red, it settled comfortably amid majestic old apple trees on a corner lot beside a dusty dirt road in historic Lexington.

Red's nine-year-old son Billy accompanied us that day. Bounding out of the car, he scampered up a climbable tree; from a perch on one of its upper limbs, he declared, "This is *it*!" We *had* to agree with Billy; it *was* the place for us. Small children gathered 'round, curious about our presence. This home's six cozy rooms with the right-colored walls, attached breezeway leading to a flagstone patio, and spacious back yard, seemed the fitting place to land. A shaded sandbox within sight of kitchen activities clinched the deal. As we signed papers we thought of our wedding day—this was a moment of long-term commitment.

RED MOVED US IN while I awaited the birth of our second child at my folks' camp by the lake. Linda Carlene began to arrive as I mounted the last of numerous photos picturing her brother from babyhood to toddler; her father had threatened to take no photos of this second offspring until I transferred the pictures of our firstborn from shoebox to photo album in some organized fashion. Linda was listening.

She came even more easily than Jonathan; a birthing mother I was meant to be. In contrast to my previous experience, a small-town doctor, a family friend, honored my wishes. No episiotomy, no spinal. He strapped an anesthetic device to my wrist, which I could sniff as needed for relief from the more excruciating moments of pain. It was over within a couple hours.

As I phoned Red about our daughter's arrival, the long-distance operator snickered, "John Raper! Ha, Jack the Ripper! Ha, Ha!" What a name with which to burden a sweet babe, I thought—perhaps not so hard on a girl as a boy—but

Mom and Pop visit our new Lexington homestead

by then I was used to dealing with such mockery. I had come to think of it as being more *their* problem than mine; after all, the *bad* word is rapist—you ignoramus. Unlike my daughter-in-law now, but like most women then, I dutifully exchanged my perfectly normal name, Allen, for a moniker tempting of taunts.

Thus began a period of over five years during which parenthood became paramount, and science, for me, lay fallow.

Chapter Seventeen

Parenting and Politics

SETTLING IN, WE LOVED our new home: its young neighborhood a couple blocks from the local school and buses, and in a bona fide town with its own sense of place, not just Boston's bedroom. Two families across each street from our corner lot became our extended family. Had it not been for them, my fear of going brain-dead as full-time parent and ex-practicing scientist could have been realized.

With seven young children among us, we had a job to do beyond parenting: We needed to improve the public schools. Charlotte Lichterman, Marge Battin, and I set to work. We identified four other like-minded residents and founded an entity called Citizens' Committee for the Lexington Public Schools (CCLP). Stay-at-home moms like us then felt compelled to improve our community. We'd been college educated and saw a need to do something of value beyond child rearing and husband care. Our husbands, too busy with their professions to help much in such matters, egged us on. While comparably educated young mothers today may want to be equally involved in communal affairs, they seldom have time; most are too busy holding down jobs outside the home.

The Lexington public school system did not then rank high on the Selectmen's list for funding. Teachers' salaries were less than adequate to support a family within the town. We hoped to persuade town leaders that improved public schooling would not only enhance the quality of life but boost the economy as well. My science-oriented mindset urged us to obtain hard evidence of this by devising and distributing a questionnaire for teachers.

Armed with data from our questionnaire, we recruited more members to the CCLP and assembled a stable of effective speakers who could meet with townsfolk, discuss our findings, and suggest possible solutions. We reached a large number of leading citizens by offering presentations at regular meetings of Rotary, Lions, Kiwanis, Knights of Columbus, D.A.R., and the League of Women Voters.

Our pitch aimed to convince a large bloc of voters of the need for raising taxes to boost teachers' salaries as a painful first step in developing a better school system that, in turn, would increase property values and stimulate local business. We had also to concentrate on identifying and supporting excellent candidates for the School Board.

This grassroots effort produced results: By the time our kids reached school age, both taxes and teachers' salaries had increased, and the School Committee began to encourage experimental collaborations with educators in nearby universities. This awakening generated enough excitement to attract the kind of innovative teachers our system needed. Gradually, Lexington gained the reputation of supporting superior public schools; good schools attracted more residents who cared about education; business flourished, and property values rose accordingly.

Ultimately our town became one of the most desirable communities in the Boston area. Property values increased ninefold from the time we'd purchased our home in 1954 to its time of sale in 1986. As with drugs that heal, however, these improvements bore unfortunate side effects: Ordinary citizens, including the teachers themselves, could no longer afford to live there—their salaries, alas, had not kept pace with inflated property values. My scientific training had not helped to resolve a problem like this.

Nor did scientific training prepare me well for all the parameters of parenthood. Nonetheless, I loved being a mother at home, helping the kids grow, laugh, and develop, being there at crucial times when important questions required response: "Why *won't* Pammy play with me?" "Why *can't* I marry my sister?" "How come Billy can hear me when I talk on this phone here and he's all the way down there in North Carolina?" or, "How does rain get made?" and "How did Soozy get her puppies?" I realize in retrospect how lucky we were to raise a family on one professor's salary and, with prudent planning, live a life of quality.

Primary parenthood

Being the youngest of a large family, I had little understanding of the excitement, frustrations, or rewards of parenting. This switch in lifestyle had its downside. Despite the camaraderie of other mothers in the neighborhood, the heady challenges of civic involvement, and the joy of making music as a member of a fine choral group, I missed science. I missed the shoptalk with Red; that link had weakened while my bonding to our offspring grew stronger than his. As much as he said he wanted children, he, also the youngest of a large family, did not take naturally to babies and inarticulate toddlers; while he loved our babies, and would protect them to the death, he preferred the company of college-age youth.

I NOTICED A DIFFERENCE in our relationship: Before parenthood, Red and I spent most of our leisure time together. To please me, he played tennis, bicycled, skied. To please him, I helped in the woodworking shop designing and building furniture. To please each other, we worked in the darkroom developing skills in art photography and, in season, sailed together. Now, no more sailing, and the joint darkroom work happened only at Christmas while making family greeting cards.

Red no longer felt compelled to play tennis, bicycle, or ski, and I participated less and less in our basement woodworking shop, while he made virtually all the furniture in our house. His creations adorn my and our children's homes even now.

After 10 years together, our marital togetherness had changed. He had his life, I had mine, and, like most married couples, we met evenings and weekends to share the responsibilities of keeping a home. The honeymoon was over.

A sample of our joint endeavor

Chapter Eighteen

Reentry Abroad

I WAS HANGING OUT the laundry one fine spring day in 1959 when Red came home early to gleefully announce he'd been awarded both a Fulbright grant and a Guggenheim fellowship for a sabbatical year at the University of Cologne. We jumped for joy, hugging each other amid the drying bedsheets. These prestigious awards, in addition to the half salary from Harvard, meant that the whole family could accompany him abroad for a full year. By then, Red had shifted research from water mold sex to mushroom sex. In collaboration with Dr. Karl Esser at the University of Cologne, he proposed to pursue an entirely new approach, using immunological techniques to investigate mating in *Schizophyllum.*

Red had studied German in college; I had not. I promptly went to the Harvard Square Coop, and purchased a short course on German conversation, complete with long-playing records and accompanying study guide.

When 14-year-old Billy Raper came as usual to spend that summer at our home, he expressed enthusiasm for going with us to Germany. Red and I agreed, with the provision that Billy should acquire some rudiments of the language in advance. Wisely or not, we thought it best to enroll the kids in German schools, Jonathan in third grade and Linda in first. We figured that at their young age, they could learn the language in the classroom without undue penalty; the equivalent of first year high school for Billy would be much tougher without some knowledge of German in advance. Besides, Billy never seemed a very eager student. He had to work hard for what he got, and, more often than not, he didn't feel like working hard. We set up a regular schedule whereby he and I would study together for a couple of hours almost every weekday throughout the summer.

Billy willingly complied at first but progress was slow; his interest waned. Playing with the neighborhood kids, all of whom were on vacation, became too tempting. Why shouldn't he be on vacation too? I gave up the struggle to keep him on track, and Red, according to his own work ethic, judged Billy's motivation for spending a year in Germany insufficient if he didn't want to work for it.

I liked Billy; we all looked forward to his visits each summer, especially Jonny for whom Bill was a good big brother. But Red and I disagreed on discipline. While my parents (primarily Mom) raised us children with firm hand, they gave us leeway to fool around (conjure, as Mom called it). Red's parents on the other hand (primarily his father) disciplined him sternly—if young John got one B grade out of all A's his father chastised him. He never could goof off from farm chores. Sunday was the only day of "rest." Card playing was not allowed, certainly not dancing, only the occasional game of checkers. Red had come a ways from that but still expected more from our kids than I did. I think this was hardest on Billy, who visited only in summer. I could try to soften but not strongly counter Red's discipline of Billy; the boy was his and Ruth's son, not mine.

Sadly, that summer was the last we saw of Billy. We all missed him when he and his mother Ruth moved from the east coast to the west coast. I learned only later, after Red's death, that Billy had changed his last name to Ruth's maiden name and regretfully failed to inform his father for fear of offense. As a young teenager Billy must have been sorely angered by Red's decision to exclude him from that year in Germany. In a letter to Billy's mother Ruth, Red expressed some remorse:

> I've been a bit uneasy about last autumn's decision about Billy's not accompanying us to Germany next year. At the time your own reaction, that you'd be somewhat happier if he didn't go, rather reinforced the decision. I suppose what bothers me is a feeling that I haven't done the right thing, or that if I have done the right thing, that I have done it badly—in such a way that Billy will be unduly disappointed or hurt, and this most certainly was not intended.

Billy was more than disappointed. Upon our return from abroad, we tried but failed to entice him to join us on a camping trip west the following summer. Billy indicated his disaffection in a sorrowful letter I received only after Red's death.

"I never really disliked my dad; I only felt resentful that I couldn't think of any good arguments with which to defend myself or my beliefs." Billy, then 29, went on to say he had been thinking about visiting his dad within the next year or two when he got the news of Red's death. By then, Bill's views on life and religion had changed, having moved much closer to his father's way of thinking. He wrote, "We had our first serious discussion on these subjects when I was 12 and a believer. Dad arbitrarily dismissed my viewpoint with a rather curt, 'You're not old enough to understand the problem. Come back in nine or ten years and we'll talk again.' Bill then explained he'd long since come to realize that his dad wasn't used to listening to 12-year-old, non-logical arguments. At the time of Red's death, he had established a career in teaching language arts to seventh and eighth graders. Unlike his father, he had learned, never to put a kid down, even when he's just clowning around. Bill ended his letter saying, "I liked my dad very, very much; the problem was that I was never able to argue with him on his level."

Bill's sentiments hurt. Had he grown up with Red as his full-time dad (or possibly come with us to Germany), he might have been able to resolve these resentments before adulthood, before his father's death. I feel remorse, but not guilt. Ours was a path taken at a point of divergence. There was no going back.

MY MOM CAME TO Germany with us instead of Billy. My dad had died in the fall of 1958 at the age of 75. He had been ill with diabetes for several years and Mom had used most of her energy taking care of him at home in the flat above his former law office, where they started, and later ended their married life together.

After Pop's death, brother Luther sensed Mom's need to get away and find new use for herself. Mom loved to travel; she read incessantly about faraway places but had never seen any outside the confines of the United States and eastern Canada. Lute had leave from the University of Massachusetts to teach International Relations at Saigon University, South Vietnam, starting the fall of 1959. He wanted to get an advanced start and show Mother Paris for a week on his way to the new assignment. At his suggestion, we asked Mom to join us afterwards for the year in Cologne, a well-intentioned plan beset with fortuitous but ultimately troubling consequences.

The leisurely six days from New York to Bremerhaven on the new liner *Bremen* provided a period of recovery from frantic preparations to transplant the Raper home, stuffed animals and all. This was my second transatlantic voyage; I *loved* the look and smell of the ocean. We sank into the sedentary ways of fellow passengers, mostly older German immigrants returning to visit their homeland 14 years after the end of World War II. We lazed in our bunks in the mornings, enjoying a "knockup" of tea served in bed, then "elevenses" of bouillon and biscuits on deck; a walkabout, lunch, deckchair reading, more walkabout, dinner, movies. The kids found other kids to play with.

Upon landing, we boarded a train to Cologne, picked up our preordered Volkswagen Bug, and settled into a prearranged second-floor apartment in one of the few buildings remaining fully intact after the devastating bombings by Allied forces in the last phases of the war. Downtown Cologne had risen from ashes around its magnificent two-spired Cathedral, but most of the outlying neighborhoods remained in relative disrepair. Commonly, big old stone buildings with ruined top floors had occupants on the ground floor or basement only.

Photographs taken just after the war showed the Cathedral standing alone in all its gothic splendor amid rubble stretching for miles in all directions. The Cathedral, while damaged, survived through techniques of precision bombing and preventive removal of stained-glass windows. As I stood before the western façade, its majestic symmetry delicately trimmed with lacelike stone moved me beyond speaking. *How can humans create such wondrous beauty—and perpetrate wars of mass destruction? Why does one culture, in a sense of self-righteousness, fight for power over another at the risk of losing everything of value to both? The survival of Cologne Cathedral is perhaps one small symbol of hope,* I thought.

Mother arrived from Paris and established herself in the formal dining room (with added bed) that could be closed off by sliding glass doors to make a private place. Red and I occupied the bigger bedroom with its massive feather bed beneath an intimidating iron crucifix hung perilously overhead. The kids had a smaller room off of ours. They played on its balcony in the company of two caged parakeets, overlooking a lovely backyard garden. Everyone had to troop through our bedroom to reach the bathroom with its large clawed bathtub, noisy gas heater, and ancient plumbing, the fragility of which was all too well demonstrated when it burst one night in our sleep and caused a flood. We had to rouse the plumbers from the corner pub.

Our landlady, Fräulein Lucy Schnitzler, made her living selling chemicals out of her first-floor apartment. She related terrifying accounts of night and day bombings in which the top floor of her building caught fire several times and everything around got badly damaged. She, her father, and the tenants would seek refuge in the subtending cellar with buckets of water to quench any fires. As they emerged between raids to check for damage, they looked toward the center of town to see if that blessed Cathedral still stood—it gave them reason to carry on.

We liked Lucy; she loved the children and became a fun-loving companion. As she spoke of the terror suffered during those awful bombings, she confessed to an added horror: the belated realization that she and her countrymen lived on the wrong side of that war. She did not hate the messengers of such destruction but felt that the Germans had brought it upon themselves: "Es ist unsere Schuld!" ("It is our guilt!") she would say. Now, with aid from the Marshall Plan, rebuilding restored hope for the nation's survival in peace.

Germany, formerly a country leading in science, had lost a whole generation of scientists during the Nazi era. Karl Esser, a surviving soldier (Herr Doktor Professor in training) welcomed us with charm and gracious consideration. He helped with anything and everything. Under the Department Head, Herr Doktor Professor Straub, Karl had arranged for an appropriately equipped lab that could accommodate both Red and me in their brand-new science building. With Mom at home to be there for the children when they weren't in school, I was delighted at the prospect of getting back into research with Red.

Red and Karl planned to explore the mode of action of the mating-type genes, using a physiological-biochemical approach involving a number of very large rabbits. While genetic understanding of the system had advanced, the molecules used for mating remained a mystery. Red and Karl reasoned that specific proteins, particularly enzymes, must play a role. They chose highly specific and sensitive methods of immunology to correlate proteins with genes. While their yearlong labors yielded data, the data were mostly of the negative sort; they identified no products of the master regulating mating-type genes but did gain some inkling of other genes being involved in the pathway to sexual union.

Schizo would not give up her secrets easily. An understanding of what those mating-type genes are doing and just how they are doing it would, as with *Achlya*'s sex hormones, have to await new technology not then available.

Despite failure to achieve this scientific goal, the year in Cologne yielded other benefits, not the least of which was revival of my scientific brain. I became fascinated with an oddball phenomenon known as disomy in *Schizophyllum*, in which the chromosomes carrying the mating-type genes fail to separate in meiosis and thus remain doubled (diploid) in an otherwise haploid genome.

The sticking together of one or more chromosomes in the late phase of sexual reproduction is not uncommon among the offspring of a fruiting stock that Hans Kniep had deposited in a Dutch culture collection back in the 1920s. An explanation for such anomalies is not clear, but it probably reflects spontaneously occurring mutations affecting the complex process of meiosis. A comparable stickiness producing triploidy of Chromosome 21 can cause Down syndrome in otherwise diploid humans.

In any case, I observed that a doubling of only the parental chromosomes carrying different versions of the mating-type genes produced a phenotype comparable to a compatible mating between two haploid strains. This suggests that the mating process is governed by interaction between products of the mating-type genes alone, and is unaffected by the dosage of genes residing elsewhere.

Why should one even think about the effects of gene dosage in mating? Well, determination of sex in humans and other animals, even the fruit fly, is critically affected by a balance in the number of gene copies on the sex chromosomes X and Y, versus the other (autosomal) chromosomes. Our observations on the genetics of sexuality in *Schizophyllum* implied a different mechanism. They led us to think it possible that mutations in the mating-type genes alone could alter sexual affinity and reveal something more about how they function—a project to be taken up at Harvard when we returned home.

Chapter Nineteen

Domestic Turmoil

THE EXCITEMENT OF LIVING abroad, taking up science again, and making new friends, was tempered somewhat by a growing dissatisfaction within the family: Having been a primary parent for the past six years, I realized the difficulty of always maintaining family contentment. Now, in a foreign place with an added family member, we started having problems.

Our youngest, Linda, did well from the start. She had an empathetic first-grade teacher, Frau Fehling, who valued her presence as an Ausländerin (alien) in an all-girl class. Lin learned how to read German at school and enlisted her willing grandmother, a former schoolteacher herself, as her tutor in English reading at home. She excelled at both endeavors, won a prize for best first-year reader in German, and made many friends who welcomed her into their homes.

Jonathan, on the other hand, fared not so well in his all-boy class of third, then fourth graders. Constantly being tested for toughness by his classmates, he often returned home with skinned knuckles and torn trousers. Somewhat stoic by nature, he refused to reveal the source of such dishevelment, not wanting parental interference.

Finally one day, while just the two of us walked together

in a nearby park, he told me he'd figured out something important about how to deal with the bullies at school: "When they try to make fun of me, I just walk away, turn around, put my feet apart, my hands on my hips (like *this*), and stare them down. Then they walk away!"

I later learned the identity of his tormentors: a pair of very large twins who took pleasure in pestering this strange boy from far away. It probably didn't help that Jonathan, when bored, had a penchant for doodling swastika-labeled Nazi fighter planes while sitting at his desk in school.

This was a lonely year for him. He had no friends or outside activities. He spent his time at home playing knights and castle on the balcony with his sister, listening to Grandma read *Robinson Crusoe, Winnie the Pooh,* and *The Swiss Family Robinson,* and reading books himself while clutching his one and only true pal, a teddy bear named "Bear." In retrospect, the little boy in grown-up Jonathan never forgave us for taking him to Germany, depositing him in a German school, and worse, leaving hopelessly worn-out Bear behind when we packed and returned to the States. While he enjoyed family travels, castle hunting nearly every Sunday, and visiting other countries, Jonathan laments even now the German experience that changed him "from a sweet, loving little boy to a youngster who no longer trusted everyone."

Mom liked taking care of the kids. They needed her nurturing; she needed them to help with the shopping, a far more arduous task abroad than at home. The day's shopping required not only some knowledge of German but also visitation of numerous specialty shops: this one for meat, that one for baked goods, another for vegetables, yet another for beverages (wine or beer was cheaper than milk). The kids provided good company and, after a couple months of schooling, the necessary knowledge of German.

Mom, lonely at times, nonetheless loved the newness of it all. She ultimately formed a pleasant friendship with another woman of comparable age, Frau Wielock, who lived upstairs

Our kids exploring Germany

amid heavy dark furniture and intimidating portraits of heroic figures including Kaiser Wilhelm, the former King of Prussia.

Red, not naturally the "happy camper" type (unless actually camping in the wilderness), had problems early on. He missed the ambiance of his Harvard labs, the graduate students left behind. He worried about their probable lack of progress without him there to supervise; a postcard sent to them, not wholly in jest, said: "Hey Guys, get the hell back to work!"

As for me, getting back into science boosted my spirits. With Mom to look after the kids, I often accompanied Red on his lecture tours as Fulbright scholar (Denmark, Finland, Italy, France) and began to feel something like his professional partner, not just a supportive wife. While being Red's wife was no small potatoes, it did not wholly satisfy. Yes, most women of the time appeared quite content as full-time homemakers and mothers, but I had always looked for something beyond that. Now, back in the lab, I began to extend my own special talents again.

WE USUALLY TRAVELED TOGETHER, but Red did some traveling on his own, specifically to England, where he visited academic colleagues and revisited Brenda, the lovely lady of his previous British experience a dozen years earlier. Brenda had not yet married; she welcomed Red's friendship, admired his sensitive hands, and tended his need for undivided attention.

As in any marriage, our relationship after a decade with kids and household duties, ceased to hold all the romance of its earlier years.

Red became troubled all over again: Was this to be a repeat of the break in his relationship with first wife Ruth? He truly did not want that, but how to resolve this strong attraction for two women within a monogamous mind-set?

I saw his distress. He could not sleep. We had to talk. We did. To his relief, I took an understanding view—*at first.* I'd had a similar experience some four years earlier, having been—now no longer—secretly entranced by another man. Furthermore, I *had been* 'the other woman' at the end of Red's marriage. I loved him so! It hurt.

Two weeks later, it hurt even worse. Now *I* couldn't sleep. Instead of the usual walk with Red before bedtime, I walked alone, lonely. One spring night, I stared at craters on the moon and wondered: *What is the use of this life? What use is any life? Our existence affects neither the moon nor anything else in the universe. What good is it?* Worry and despair overshadowed all the enthusiasm I'd had for this year abroad.

I wrote a poem:

> It is not known from whence we came
> or why.
> Nor is there one who truly knows
> save God.
> But is there God beyond our need
> for certainty and comfort?

By spring, the torment intensified. Red became so troubled he simply could not handle the continued presence of my mom within the household. He'd always been somewhat irritated by her habit of rhetoric, her well-meaning, simplistic queries about his work, and her frequent references to the Allen family, of which she *assumed* he was a full-fledged member. I'd been surprised at the length of his patience, but now, on the verge of a breakdown, he beseeched me to ask Mom to leave early and forgo her longed-for voyage home with us by transatlantic liner. Our family was falling apart! Knowing that Lute was returning soon from Saigon, I wrote to him for help.

Lute came through in a letter to Mom inviting her to meet him in Greece, tour the Greek islands by boat, then visit Egypt before flying with him back home to the States.

Mom, nonetheless, expressed a preference to stay with us through the summer as planned. We then had a tearful heart-to-heart talk. Declaring my love for both her and Red, I had to explain Red's true state of mind. We had reached a point where I had to choose between him and her.

Mom did understand. She'd always believed a woman should stand by her man. She agreed to meet Lute on his way home in April. She never got the ocean voyage she craved, but the sum of her journeys that year—Paris, Cologne, Scandinavia, England, Holland, Greece, Egypt—helped her embark on a life of widowhood. I'm glad she was there for the kids and for me. A few months later, so was Red.

THE FIRST ALL-GERMAN BOTANICAL Congress was held in Halle, East Germany, behind the "Iron Curtain." That wall, built by communist Russia after World War II, imposed a serious travel barrier between the Russian-dominated East and the Anglo-dominated West. Yet in the spring of 1960, plant scientists on both sides of the border somehow arranged a meeting that necessitated travel from West to East. Red was invited to present a talk (auf Deutsch). Thinking his personal dilemma with respect to Brenda might get resolved by including her on a week-long trip with us through Switzerland prior to that meeting, he invited Brenda to join us. "Would that be okay with you?" he later asked. I acquiesced.

In accordance with this dubious sense of logic, we met Brenda at the Cologne airport and headed south—I in the back seat with the kids, Brenda in front. Red seemed happier than he'd been for weeks. Brenda was politely appreciative; I hid resentment with a modicum of cheerfulness while the kids remained clueless. I recall an evening's stroll down an elegant shop-filled street in Lucerne, Red with his arm crooked in mine on his left, Brenda's on his right—the kids trailing—when he proclaimed pleasure at being the escort of his two favorite gals in all the world. I knew then for sure that this holiday with Brenda would not resolve anything.

After parting with Brenda in Geneva, we headed northeast toward Bavaria and a magical tour of the Zugspitze, one of the more magnificent Alps accessible by Seilbahn. The gondola lifted us up over lesser peaks that glistened like inverted icicles. It let us down at the German entrance to a long ice-encrusted tunnel that led past the Austrian border out into a sunlit snow bowl beset with weaving skiers. We viewed the enchanting scene from the comfort of a half-timbered inn, tucked into the slope overlooking the bowl. After a welcome respite, we descended to base, spirits somewhat renewed.

A difference between West and East became apparent as soon as we crossed the border into East Germany. Halle looked gray and glum, the people quite shabbily dressed. Our hotel reeked with the odor of a recent fire and offered few of the amenities to which we had become accustomed in the West.

Linda came down with a fever and rash on our second day. She recovered after doses of penicillin, prescribed free of charge, and obtained from an archaic-looking pharmacy with leeches displayed in its window. I've wondered since whether that episode might have had anything to do with her having developed multiple sclerosis 20 years later—probably not.

Red's talk in German went fine, he having been lubricated ahead of time by a mug of good German beer. But the very next day I ran into trouble trying to photograph a pompous-looking building strung with a humongous banner proclaiming "AMERIKA, HÄNDE WEG VON CUBA", a reference to the aborted "Bay of Pigs" invasion of Cuba during the Kennedy administration. As I raised camera to eye, ready to click, the building's front door opened and out spilled six uniformed figures, three to either side, marching toward me in perfect formation.

I looked back longingly as Red sat helplessly in our parked Bug with the kids. The six guards ushered me into the building and marched me through a long hallway aligned on both sides with soldiers at strict attention. I felt chagrined, dressed as I was in tourist garb: Greek skirt, Norwegian sweater, bouncing camera slung around my neck. They whisked me away to a back room within the depths of that cavernous edifice. An officious suit, speaking German, began the interrogation. I stood tall and mustered my best poor German: "Was bedeutet's? Ich bin die Frau von Herr Doktor Professor Raper aus Harvard Universität, aus die Vereinigen Staaten von Amerika! Wir zu Halle für das Deutsche Botanische Konferenz kommen. Warum bin ich hier?" In short, 'What is the meaning of this?' My interrogator revealed the purpose of this building as Russian Army Headquarters for the entire region—unauthorized photographs "verboten!"

I explained that I had not clicked the shutter, which was true. Nonetheless, he disappeared with my camera and, after some minutes, returned it to me without film. All those photos of the beautiful Zugspitze gone! In answer to my query as to why I should not photograph the Russian Headquarters, I was informed that unofficial photographs of military installations in my country are also "verboten." He let me go without further incident. This reality check on the Cold War put a damper on our spirits. We had thought to make a side trip to Leipzig, where Bach had performed over two centuries earlier, but decided not to risk it.

While stopping by the University of Jena on our way back toward the border, we visited an honored colleague, Dr. Manfried Girbhardt, and told him the story of my capture. He was not amused. He told of another recent visitor from the West who had experienced a similar incident with different consequences; that person was likewise detained but incarcerated in prison for some length of time thereafter.

SETTLING DOWN TO THE three remaining months in Cologne, we managed the work well enough—the kids had adjusted to their school routines—but, as I had suspected, the Brenda business persisted. Red continued to suffer insomnia and I suffered accordingly, feeling increasing need for his amorous attentions. For the second time in my life of loving him, I worried about his stability. My anxiety peaked in Rouen where we stopped overnight on our way to Le Havre, the port of boarding for the planned voyage home on the French liner *La Flandre.*

Red, in desperation, plotted a last-ditch stand with Brenda. He would leave Rouen for London that evening, spend the next day there, then meet our boat as it docked at Southampton. I was left with the task of explaining his sudden absence to our children. After weeping half the night, my stamina seemed barely adequate for getting us and all our belongings to the boat. My feelings then must have resembled Ruth's when Red defected to me over a decade earlier.

Our boat pulled into Southampton harbor at sunset, August 9th—our 12th wedding anniversary. In rejoining Red I'd hoped for warm greetings and some acknowledgment of this special day, but he showed up exhausted, went straight to our stateroom, and fell promptly asleep. I had neither given him the cufflinks I'd bought as a gift nor found out what happened in his visit with Brenda.

I sat alone on the upper deck until late that night, watching the stars, hearing the music to which others danced, and hoping in vain that Red would come find me and all would be well. I knew he'd have a story to tell. I only hoped his profound state of fatigue leading to such irrepressible sleep signaled some kind of favorable resolution.

On the next night at sea, once the kids got tucked in, we sat together in deck chairs and talked. Red, now rested, confessed to forgetting our anniversary but considered that minor. He spoke with relief about Brenda. She had not encouraged him. They had talked through the night, setting it straight that she would be no more than a sisterly friend who respected us both, the sanctity of our marriage, and the importance of our role as parents. She helped Red to deny any future with her. I never learned the extent of Brenda's feelings for Red, but if her love for him was anything like mine, her noble restraint exceeded mine when Red was married to Ruth.

Brenda later married a charming intellectual who required ample attention. She complied, even washing and ironing their bed linen daily, as per his needs. We became friends. I thank her for the way she said goodbye to Red in the early morning hours of August 9th, 1961. Red had been in crisis. She helped him cope.

I asked myself: *Is such trouble common in long-term marriages?*

In our case I wondered if my previous change in lifestyle, being out of science for so many years, contributed to our dilemma. Red always valued our talk about science; in fact he seemed to depend on it. The six years when I could not share that bond as intensely as before had taken its toll on our marriage, just as it did in Red's first marriage when he could not discuss his secret research at the Manhattan Project with Ruth and she had become the primary parent.

We both needed time to heal. Now, as we entered another phase of our union, something had to change.

Chapter Twenty

Resolution

FOLLOWING THE YEAR ABROAD it was a time for me to make decisions about my future. Should I resume the role of social activist and perhaps aim toward political office? Lexington seemed to have done well in my absence even though a very personable plumber who knew little about running a school system had been elected to the School Board while I was away. Should I aim to become a writer (a bold but lonesome thought) or keep doing science? I had gotten used to a home without phone calls urging my service in one capacity or another. Perhaps the best choice would be to do something part-time at home, to always *be* there for the kids when they needed me. But I liked getting out of the house into an academic workplace, and I liked the research I had started in Germany. Deep down, I must have realized that continuing with the research would benefit my marriage.

After a week's deliberation, before most of my former activist buddies knew I was back in town, I chose part-time work in Red's lab. The family agreed to the plan, Red especially—thinking perhaps that a shared life in science might revive the romance of our courtship years. I would ride in with him in the morning and return by bus in the afternoon in time for the kids' arrival from school. The pay would be a pittance, but here was a chance to resume a career I knew and loved. Thus began a disciplined life of diverse responsibilities—something called *women's work.*

Immersed in the heady atmosphere of the Harvard Bio Labs, I experienced a kind of excitement not previously known to me. An official position as tech allowed me to spend all my working hours doing the research I desired, while relating directly to an array of talented graduate students, undergrads, postdocs, visiting professors, and other techs, all under the aegis of my husband, who had to spend his days keeping the whole enterprise going.

I felt rather sorry for Red, chained as he was to his desk, preparing lectures; working on grant proposals; planning scientific conferences; writing papers and letters of recommendation; phoning. When not engaged in those activities, he'd be off to some committee meeting or teaching a class. It was only later, when he reached sufficient status as chairman of the Graduate Committee, that he merited a full-time secretary to answer the phone, type his memos, and generally shield him from constant interruptions.

Red, like most other professors tenured in science, had no time for hands-on research—the very activity that drew Harvard's attention in the first place. Red sometimes sighed, bemoaning the suspicion that the Peter Principle (promotion beyond capability) applied to himself. I was to learn something of this firsthand when I ran my own lab many years later.

My research project of choice was to get some idea of how many genes beyond the master regulating mating-type genes might be involved during the process of mating in *Schizophyllum.* We knew from the immunological experiments that a series of proteins had to be involved, but where are the genes encoding those proteins? How many are there?

A good starting point would be to examine spontaneously emerging mutants that disrupt the process of fertilization.

Using knowledge developed by Haig Papazian, I chose to select the mutants from a mating of two individuals incompatible for the *A* mating-type genes (identical *A's*) but compatible for the *B* mating-type genes (different *B's*). Such a mating produces a depressed phenotype against which mutants disrupting the mating process can be seen as healthy white fluffy growths.

Unlike fertilization in humans, where the sperm nucleus travels on a one-way street to join and fuse with the egg nucleus, fertilizing nuclei of the two fungal mates compatible for *B* travel on a two-way street; they donate and migrate reciprocally throughout each other's entire body (the mycelial colonies), and do not fuse until much later in specialized cells—they might be thought of as egg cells—on the gilled surface of the mushroom.

This process takes a lot of energy away from normal growth; the mated colonies look rather sad and depressed. The human process takes energy too, but its prelude is more often exhilarating than depressing. If the fungal fertilizing process is interrupted by some mutation, normal-looking growth will be restored, generating healthy-looking fluffy colonies against that depressed background.

I isolated and analyzed hundreds of such colonies, and found over 80 mutants of a dozen different phenotypes that block the mating process but do not interfere with normal, non-sexual (vegetative) growth. Wow, mating, even in this lowly organism, must be exceedingly complex, run by a myriad of genes. I, and others in the lab, worked on locating them. They mapped all over the genome, some linked to the mating-type loci, many not.

The normal, wild type versions of these mutants are expressed only when the *A* and *B* mating-type genes get turned on by coming together in a compatible mating. Any set of genes, which by virtue of a difference can switch on expression of so many other genes, must be interesting. How many switches are there? What is their nature? What kind of products do they encode? Are they similar to genes found in other organisms or are they unique to *Schizophyllum* and perhaps other related species? How do we get at these questions? This would be taken up later.

Meanwhile, settling back into our home, sharing science and parenthood, Red and I affirmed our need for each other. Our bonds grew stronger in a second decade of marriage.

Chapter Twenty-one

Females in Science

THE ATMOSPHERE AT HARVARD resembled none other in my experience. We lived in an era of change. The established professors behaved as gentlemen of the old school, dressing in neatly pressed trousers, jackets, and ties; they held doors open for ladies and made time for friendly talk. The younger set wore jeans and sports shirts. They constantly hurried from here to there, and spared little time for polite conversation. We, falling somewhere in between the old and the young, related to both groups. Red, who preferred informality with his own students, could not, however, manage to address his former professors on a first-name basis even though they called him John or Red. I, as a female becoming bolder, had no such problem.

I came to understand why the older lot of professors had made it to the top. Tall, handsome Professor Kenneth Thimann, for example, distinguished himself not only as plant physiologist who studied growth hormones but also as poet, pianist, composer, and gardener. Conversations with Ken and his wife Ann, a weaver of fine fabrics, covered a wide range of topics from art, music, politics, and science to child rearing and social injustices. Their aesthetic senses suited ours. Whenever they visited our home Ken always noticed anything new or unusual.

As previously mentioned, Red had a hobby of making furniture in the basement. When he finished each piece (a chair, a couch, a bureau) he would bring it upstairs and put it in its designated place. Others seeing this custom-built furniture for the first time might comment or not, but Ken, in curiosity, would bend his long frame, crawl down on the floor, and examine the nature of its joints, the quality of its wood, the fineness of its finish.

On one occasion, Red had just made a discreet clock for me to have in the living room. This was after I stopped sharing shop work with him. Its mechanism was that of an old electric clock mounted behind the door of a white Cape Cod style cupboard. The clock had no face. Its hands were replaced with two black rectangular pieces of metal, one shorter than the other. When at right angle, they resembled the cupboard door's two black hinges. This simulated hinge changed gradually with time. One evening when the Thimanns visited along with two other couples, I noted it was time to check the roast. Ken, seeing I had no watch, asked how I knew the time. "By the clock in the living room," I said. Our guests scanned the room for any evidence of such, but Ken was the first to spot the misplaced hinge that moved with the minutes. I suppose one of the reasons Harvard hired him was because of his remarkable sense of curiosity and superior powers of observation.

All the more established Harvard professors had well educated stay-at-home wives whose primary focus was on husband and progeny. Ken Thimann's wife Ann, for instance, raised three girls and hand-loomed beautiful woolens. Paleontologist Al Roemer's wife Ruth kept a grand house in order, prepared the meals, organized dinner parties, raised two boys, and made all the travel arrangements for Al's worldwide lecture tours. As their children became more independent, Ruth turned to political activism, did a great deal of gardening, and managed their inherited summer place on a large acreage near Amherst as well. Olive Wetmore, wife of botanist Ralph, had taught school and wanted to go on to graduate work but never seemed to manage it—her two boys kept her at home.

The "young Turks," on the other hand, made little time for socializing. When they did, they included grad students in informal gatherings, pop music loud, lights down low. Their wives, if they had any, were poets, artists, or scientists.

Soon after Jim Watson (of double helix fame) joined the Department, I recall a crowded party at his rather small apartment just off Harvard Square. Some folks danced but most tried to converse over booming thumps from amplifying speakers. At one point, Ursula Goodenough (a graduate student who later became a junior faculty member and my graduate advisor) and I were sitting on a couch when Jim plunked down between us. As he ogled the dancers he lamented the lack of a girlfriend. "Well," said Ursula, "Maybe you'd get one if you'd just take a bath!" We were all perspiring in such close quarters and pretty much "in the cups" as well. Ursula, even before the women's movement, was typically outspoken.

Tension prevailed between young and old. The newer recruits, at the crest of modernity, bore some resentment toward the more established members of the Department who held the reins of power and sat in judgment of them. The older members, having already made a place in science but not yet ready to quit, felt somewhat challenged by the arrogant young.

Like the young opportunists who had jumped on the bandwagon of biochemical genetics during the 1940s, young biologists of the 1960s and 1970s joined the ranks of molecular genetics after the modeling of DNA by Watson and Crick. Both young and old bore woundable egos, a common denominator of high achievers.

The conflict at one point reached such depths that one of the younger members of the Department refused to speak directly to one of the older professors—communication between the two had to be conducted through a third party. One faculty meeting broke up over the younger member's contention that he deserved three times as much space as the older member because his research was three times more important, a tactless assertion perhaps not without merit, since that younger member later achieved the distinction of Nobel Laureate. Some of those brash "young Turks" behaved like rebellious adolescents disdainful of their elders' presence. They signaled a trend of the 1960s fraught with protests against the interminable Vietnam War, race riots, and assassination of three important leaders.

I noticed another trend: a changing attitude toward females in science. Just before Red joined the faculty in 1954, the female Radcliffe undergrads were taught in separate classes. Soon after Red's arrival, females and males became integrated within the classroom. The tenor of those times is expressed in a letter Red received in the early 1950s from Nobel Laureate George Beadle, who had gone back to Cal Tech after having left his professorship at Harvard:

> Here is one of our fellowship folders plus a sheet of supplementary information about graduate work in biology at Cal Tech. If you know of students who might be interested, will you please pass the information along to them? Believe it or not, we now admit women for graduate work. The Dean says they have to be very good, though!

At the time, I did not find such sentiments surprising.

The old boys' network prevailed in the 1960s, and the old boys were only beginning to reach out for girls in the more advanced realms of academia. Harvard may have exceeded Cal Tech on that score, but not by much. I recall a conversation at our tea table with the immunologist Alvin Pappenheimer. He questioned the wisdom of training female graduate students in view of the probability that they would just get married and follow their husbands wherever they had to go for *their* careers.

Even though some women numbered among junior faculty at the time, no females had yet reached tenured positions in biology. A pattern was emerging: the ratio of genders majoring in biology as undergraduates approached equality, but that ratio became increasingly skewed in favor of males as the position became more competitive, a phenomenon still evident 40 years later, although to a lesser extent.

For example, in the Faculty of Arts and Sciences at Harvard University in 2005, the proportion of women to men at all but one level was relatively low: 42% bachelors, 29% doctoral, 20% postdoctoral; 76% tenure track, and 10% tenured. This must have been a time for examining a high proportion of women in an effort to determine their fitness for promotion to tenure. It no doubt reflects some effort to address gender inequities.

The statistics for Harvard Medical School comprised an unbroken line: 51% doctoral, 37% postdoctoral; 40% tenure track, and 16% tenured. The School of Public Health, somewhat less biased, still had a "leaky pipeline" as female professors reached for the tenured level.

Of 15 grad students Red had in his lab during the 1950s through mid-1970s, four were female. The ratio of female to male among postdocs and visiting professors was 3:10. Most of the techs were of female gender.

Red's female graduate students did not fit Pappenheimer's image of acquiescing women. For example, Siggie Newhauser, having come from the Deep South, knew the Bible and accordingly emerged indignant over females cast as mothers and wives, not leaders in the Christian faith. Her college transcript showed *A* in every subject except Bible Studies, in which she rebelliously made a bare passing grade, a record that favorably impressed Red with his bias against "Jesus Jumpers."

Furthermore, Siggie told everyone who'd listen: "As the eldest of 13 children, I've had all the mothering I can take!" She truly sought sterilization as a means of precluding any more responsibility toward babies and children, but doctors refused to accommodate. A strikingly handsome young blond, she dressed provocatively yet deplored unwanted advances from strange males as she walked down the streets in her female way. We shook our heads, half smiling at such complaints.

Siggie's choice of research for a thesis project fit her independent spirit: Instead of *Schizophyllum,* she chose to study a strange mushroom-producing fungus called *Cyphellopsis anomala* with a mystifying sex life. It could go through the sexual cycle not only all by itself but by mating with others as well (shades of *Paramecium*). When given a choice, it preferred the latter option.

Dr. Siggie got an academic job in the Southwest after graduate work. She had a steady boyfriend, but whether or not she married and changed her mind about mothering, I do not know.

Carol Deppie, another strong, good-looking female grad student, had run away from home and a stern, military father at the tender age of 15. She hadn't looked back since. To the disgruntlement of her male peers, Carol flaunted her smarts and thereby gained what they thought was undue attention from their common mentor, Red. I began to realize how much grad students resemble siblings in their competition for favoritism from a shared figure of authority.

Carol's fellow students judged her relatively long on speculation and short on productive results. Upon completing her PhD, Carol landed a promising position at a prestigious state university that for some reason lasted only a few years—perhaps she was *too* outspoken. The last I heard, she was living a hippie life peddling homemade soap.

Judy Stamberg, a small brunette with sharp facial features, was perhaps the cleverest of the lot. She completed an excellent thesis on genetic factors influencing extremely variable recombination between the linked mating-type loci in *Schizophyllum,* and extended those studies in academia at the University of Tel Aviv in Israel. Judy married an Israeli, had two children, moved back to the USA, and studied disease-causing genes in humans at a medical facility in St. Louis. She then suffered widowhood, remarried, and continues in her profession to this day. Judy seems to have been able to do it all, as so many women fear they cannot.

The gutsiest of all was a tiny Mexican woman of Russian descent named Celia DuBovoy. She hobbled purposefully on crutches into Red's office one fine spring day and beseeched him to accept her as a graduate student. "Why do you want to join my lab?" he asked. She replied in husky voice, "I want to study with you because I admire your work!" Red inquired of Celia's background: As an advanced doctoral student at the University of Mexico studying the hallucinogenic fungus *Psilocybe,* Celia proclaimed her desire to extend that work with the kind of genetic analyses Red was conducting on *Schizophyllum.* Red suggested she finish her thesis first, then consider a postdoctoral position in his lab. "No, that will not do," said Celia. "I want to come now!"

Celia suffered from a rare genetic condition called Gaucher's disease, for which she was being treated at Boston Children's Hospital. She had to travel from her home in Mexico City to Boston several times a year for treatment, some of which was experimental. The disease is caused by a genetic defect that interferes with the proper metabolism of fatty acids and elicits myriad symptoms, including enlarged spleen, brittle bones, skin discoloration, and distorted voice. Celia's case did not fit the norm, primarily because she had survived longer than most victims. Red could not help but admire her determination; he acquiesced even though he had doubts about the extent to which her physical infirmities would interfere with the work.

The graduate committee turned Celia down on her first try. The committee thought it made little sense for her to start a doctoral program all over again at a very different institution while being so close to obtaining a PhD at the University of Mexico. Nonetheless, Celia persisted; she desperately wanted to learn genetic techniques and apply them to *Psilocybe* right away. She wanted to explore the genetic basis of the hallucinogen that induces an ecstatic state when one eats the mushrooms of this fungus, a practice long known among Mexican Indians for eliciting mystical effects during magical ceremonies.

In response to Celia's appeal, Red explored the possibility of seeking a one-year appointment for her to join the lab as Research Fellow and learn about our way of studying *Schizophyllum.* The appointment came through eventually, and Celia arrived in the cold of February 1969. She occupied a bench opposite mine. I helped as much as possible by showing her where everything was and explaining our methods of making and characterizing mutants.

Well, in less than a year, Celia turned sullen, thinking I was telling her more than she needed to know. Perhaps I was. We faced a problem in our attempts to communicate. She had the habit of starting a conversation in the middle of a thought process, thus necessitating reconstruction of what she might have been thinking before she started speaking. Sometimes I guessed wrong—a frustration for both of us. In any case, Celia learned quickly and did seem ready for greater independence. Red solved the problem by putting Celia in another room with a congenial visiting professor from Czechoslovakia, Jan Necasek, who never ruffled anyone's feathers. Celia and I stayed on good terms while functioning at a greater distance from one another. Remembering this lesson, I applied it successfully in my own lab some two decades later.

With Red's support, Celia applied once again to the graduate program at Harvard and gained acceptance starting early in 1970. She got hooked on *Schizophyllum* and never returned to *Psilocybe.* She chose to extend the work I had started on the characterization of genes modifying sexual development. She focused on a cluster of nine such genes that block migration of fertilizing nuclei. Through complementation tests, Celia showed that several genes function differentially and cannot recombine, while others recombine but have apparent identical function. In the absence of molecular cloning techniques, the meaning of this curious finding could not be resolved beyond speculation.

Celia fulfilled the requirements for a PhD, but not before a troubling incident at an exceptionally stressful time in her life. While preparing for her qualifying exams, she bore the shocking news of her mother's sudden death in a car accident. The qualifying exam is a milestone feared by all who have to take it. Failure means dismissal, usually with a master's degree only. Then, at Harvard, the student faced a committee of examining professors who questioned the candidate thoroughly on thesis research and related topics.

The questioning took place around a long polished table in the middle of a room lined with framed photographs of stern-looking full professors whose eyes seemed to penetrate no matter where one sat. One never knew in advance just what questions would be asked or how they might be answered. The exam could go on for hours. Students disaffectedly dubbed this room of inquisition "The Gas Chamber."

Sadly distressed by her mother's death, Celia arrived at the lab the day of her exam and found what she thought was a black eight ball sitting on her desk. She got through the exam all right but suffered a breakdown the next day. Bob Ullrich, a fellow graduate student more than twice Celia's size, happened to be the first person she encountered that morning. Flailing her crutches, she chased Bob around the tea table, shouting, "You put me behind the eight ball just before my exam!" Of course he hadn't—Bob had been especially nice to Celia from the time of her arrival.

We tried to figure out what had happened. No one in the lab played pool, but some played squash. A black squash ball could have fallen out of someone's pocket and the janitor may have picked it up and put it on Celia's desk. In her state of distress, Celia could have mistaken the squash ball for a number eight pool ball, maliciously placed there as ill omen.

Upon learning of this incident, Red lost patience. He counseled Celia in no uncertain terms: She "must not disrupt the workings of the group by physically attacking anyone for any reason whatsoever." Red called Celia's doctors and learned only then that a symptom of her illness was periodic mental lapses, often coinciding with her menstrual periods. By mutual agreement, Celia stayed home when such lapses occurred thereafter.

Fortunately for us, Celia played an indirect but essential role in bringing the research with *Schizophyllum* into the molecular age—and keeping it alive to the present time. More of that story later.

Chapter Twenty-two

Tea Table Talk

RED'S LAB COMPLEX INCLUDED a room with sink, stove, refrigerator, table, and chalkboard, accommodating lively discussions over tea, coffee, and brown-bag lunches. Margit Bensen, custodian of preparatory materials for our research, and resident counselor of troubled graduate students, put the kettle on around 10:30 in the morning for those who wanted a coffee or tea break; brown-bag lunch happened around noon to 1:00 PM, and afternoon tea at about 3:30 PM

I don't know a lab group today that devotes as much time or facility to informal discussions around a tea table, but the effort then was clearly productive. Freewheeling talk naturally centered on science but often included politics, social problems, music, gossip, or art. Tumultuous events of the 1960s and early 1970s encompassing threats of nuclear warfare, our futile engagement in Vietnam, violent uprisings for civil rights, the successive assassinations of President John F. Kennedy, Martin Luther King, and J. F. K.'s younger brother Bobby, could not be ignored. These topics of despair sometimes elicited vigorous argument or proactive efforts.

For example, Abe Flexer—a handsome, blond grad student going on postdoc in the lab—and I grew so concerned about discontent in the Black ghettos that we helped found an organization called Fund for Urban Negro Development (acronym FUND). FUND aimed to raise money and develop a business-skills bank available to the Boston Black community in Roxbury, no strings attached. Economic depression in that community stemmed partly from lack of local control and the prevalence of exploitive absentee landlords.

We hoped to encourage local business ownership by providing loans and expertise to be managed primarily by local leaders. FUND attracted like-minded citizens in the Boston area who donated money and entrepreneurial experience as wanted and needed. We adapted an old idea from the prominent civil rights leader Bayard Rustin, who said that if every concerned citizen tithed ten percent of the amount each paid in federal income tax to building negro communities, we could begin to address the inequities those communities had suffered since emancipation.

Well, of course tithing never did sell, but something got started. FUND identified a strong, trustworthy leader in Roxbury who had helped establish a locally owned bank and who agreed to work with our organization. His name was Mel King (he later ran unsuccessfully for mayor of Boston). I never did see his eyes; whenever we met, he and his cohorts wore impenetrable dark glasses, hiding any sense of emotion that might have been aroused through our discussions—it was the beginning of the Black is Beautiful movement.

All donated funds went into an account at the Roxbury bank under the management of King and his associates. An array of willing folks who had successfully run small businesses such as chicken takeouts, shoe stores, and book stores, also offered free advice to any who requested it.

I honestly cannot assess the ultimate effectiveness of such efforts, but we felt we had to try. As drawn as I was to political activism, I realized once again that any long-term involvement of this sort is not one of my special strengths. I came away feeling nowhere near as forceful as Abe, for example, at persuading others to part with their time and money for a cause. The agony of my inadequacies kept me awake at night, while Abe claimed his sleep suffered little from such activities. Whether because of inherent capabilities or societal norms, the men seemed to deliver the more compelling speeches. Perhaps as female I was not raised to feel entirely comfortable about promoting a cause with passion before an audience of strangers. This lack of confidence would plague me later while speaking publicly, even about my own science. It was a problem I would have to overcome.

VIRTUALLY ALL MEMBERS OF the lab agreed on our nation's need to get out of Vietnam. Red and I protested by defying the law and spending a sleepless night on the Lexington Battle Green. We joined hundreds of other citizens in support of the Vietnam Veterans Against the War, who wanted to bivouac there, on their famous march from the Rude Bridge in Concord to Bunker Hill in Charlestown over Memorial Day weekend 1971.

The Lexington Board of Selectmen intractably refused permission for protesting veterans or anyone else to sully this sacred ground by their presence. They cited Section 25 of the town's old bylaws, stating, among other things, "No person shall engage or take part in any game, sport, picnic, or performance on the Battle Green without the written permission of the Board of Selectmen or other Board having charge and control thereof." Like the statue of Minute Man Captain Parker defending that Green at its apex, the Selectmen stood fast, adamant in their refusal to grant the veterans permission to occupy its premises.

We heard John Kerry speak, and stayed our ground. If the veterans were to be arrested, we would be too. Along about midnight, a line of yellow school buses pulled up alongside the Green. Local police and state troopers, menacingly outfitted in blue battle dress complete with face masks and billy clubs, emerged from the buses and arrested as many as the buses would hold. They never got to half of us, including Red and me.

We spent the whole night there. Daughter Linda, then a teenager, confessed once grown that she took advantage of our absence by spending that night with her boyfriend. Unlike my idol Eleanor Roosevelt, I suppose I should have been home parenting instead of politicking. Well, thank Gawd our daughter knew more about birth control than I did at her age!

Despite such demonstrations, the Vietnam War went on for four more years. The Battle Green incident, however, prompted a strong desire to revise the Lexington Board of Selectmen. We conducted a massive door-to-door campaign and elected candidates who could function more flexibly and less conservatively than the Board of 1971. Among the new selectmen was our neighbor Marge Battin who, as I had considered doing, chose a career in government.

Marge's skill in her role as Selectman culminated in her election as the first female president of the Massachusetts Selectmen Association. She later occupied the powerful office of Town Meeting Moderator for over two decades. Marge achieved far more success as a politician than I ever would have. Her initial working tactic was to appear flirtatiously attractive, wearing lipstick and high heels at every male-dominated meeting, thereby discouraging rough male tactics laced with coarse male humor—a pose I fear I could not have mastered. Just as I knew that Red made the right decision to choose science over trumpeting, I knew then that I had made the right decision to choose science over politics.

BEYOND POLITICS, SCIENCE TALK around the lab's tea table often led to productive ideas about the research. One idea in particular sparked my interest: how to explore the evolutionary origin of so many different mating types in *Schizophyllum*. Since a mutational approach produced such a gold mine of information on the multitude of genes involved in the fertilization pathway, it seemed reasonable to try a similar approach with the master switch genes themselves. We might be able to generate a new mating type in the lab and thus get a clue as to how it happened in nature.

Yair Parag, a determined graduate student from Israel, isolated the first mating-type gene mutant in one of the two *B* mating-type loci, a damnably difficult feat others had tried and abandoned. The trick was to devise an effective system for spotting and selecting such rare mutants. His complicated scheme, different from the one I had used to find all those mutants in the fertilization pathway, yielded a mutant that surprisingly switched on the *B* pathway all by itself in the absence of mating with a compatible partner. When mated to a partner carrying any of the other *B* mating-type variants known to exist in nature, including the progenitor *B* type, it switched on the *B* pathway of development in the mated partner.

This bizarre result astonished us all; we had expected Yair to generate new *B* mating types comparable to, but perhaps different from, all the *B* mating types found in nature. Well, Yair's mutants were different all right, but not like the other natural types. This unexpected outcome prompted passionate group discussions (hindered somewhat by Yair's thick Hebrew accent and animated efforts to achieve some understanding with English-speaking Americans by just speaking louder). If a new mating type can't be generated in one step, how many steps might it take? What do we do next to try and find out?

I don't recall just who came up with what hypotheses and how such might be tested, but one seemed the best bet to pursue: What if it took more than one mutational event to produce a new *B* mating type comparable to those derived in nature? Might we use Yair's mutants to try and make another mutation in the same locus, and see if we could generate a secondary mutant in which the *B* pathway is turned from *on* to *off,* that retains compatibility with all the other *B* types? This was to be my next project.

Such experiments require a lot of tedious manipulations. Fortunately, Eva Maria Schneider, a tall, capable, energetic blond woman, had just joined our group as lab technician. True to her German heritage, she did everything in orderly fashion and seldom made mistakes (at least any she deigned to admit). Eva loved working with her hands; she seemed incapable of sitting still. If she arrived in the morning before anyone else, she would start doing the janitor's job, washing windows, sweeping the floor, both of which, in her judgment, always needed cleaning. Obviously Eva's work ethic fit our

Eva looking for mutants

task of finding the mutants. It was my job to keep her busy and think it through.

Two years' of work produced 93 secondary mutants with a wide range of phenotypes, but nary a one that fit the exact description of a new and different mating type comparable to those found in nature.

When the bulk of these results on secondary mutants emerged, Red was away in Israel on his second sabbatical leave. The kids and I were to accompany him for six months that fall of 1967—but the "Six-Day War" intervened. While the war ended on June 11 with Israel clearly triumphant, its disruptive effect altered Red's plans for a sabbatical at Hebrew University in Jerusalem. The longer stay for all of us could not be arranged. Red opted for three months alone, mostly lecturing and writing in consultation with former students Yair Parag and Giora Simchen at Hebrew University, Jerusalem.

It was the longest period of separation since our wedding. Our mutual attachment had deepened since the last sabbatical. Working together as scientists had no doubt helped us heal from the Brenda affair. We respected one another in our complementary roles and took pleasure in functioning as a team. Making love was really good again.

Now, with Red in Israel and me managing the home front, we exchanged letters daily. Red wrote about his teaching and lecturing, his adventures with Yair and Giora exploring Israel's newly gained territories, and his loneliness for home. I reported our work at the lab, the kids' doings, the little trips we took together, and how much we missed him.

About two months into this separate existence, I noted a drifting away from those little routines the children and I had shared with Red: We had supper according to *our* schedules. We ate more of what pleased *us* and less of what pleased *him.* We went to the seashore on a whim of the moment, popped out to Amherst to visit brother Luther, or drove to New York City for the fun of it. I began to stray toward the adventurous habits of my Dad, who when I was a child, delighted in surprising the entire family by waking up early, packing us into the

Franklin car, and whisking us off to some faraway place. It might be to New York City or Montreal or just getting lost on winding back roads through the hinterlands of New York state or Vermont, with the promise of good eating at some fine restaurant he always managed to find at the end. We'd travel as long as the money in his wallet held out. I must have inherited some of that spirit, and it was just now emerging in Red's absence.

I recall writing Red that he'd better hurry home, because the custom of making small daily adjustments in deference to each other's preferences was slipping away. He too was adapting to different routines far away from home.

As much as I missed Red, those three months meant a period of growth for me in the lab. It reminded me of the time in my first job at Argonne when, in absence of the boss, I'd had to make my own decisions.

Red writing

All those *B* mutants that Eva and I generated cried out for interpretation. I got the idea of arranging them in some orderly fashion from the least impaired to the most, starting with a mutant that behaved *almost* like a new natural mating type, to null mutants that could not mate at all. In matching the mutants of increasing impairment with their ability to interact with the many natural *B* types, I spotted a pattern that could tell us something about the complexity of the genes involved and how their many different versions might relate to one another. The variety of phenotypes suggested at least two kinds of genes within each of the two *B* loci: one for donating and one for accepting fertilizing nuclei. They also implied a switch of some kind with a turn-on/turn-off function.

When I explained my ideas to Yigal, a grad student at the time, he remarked in astonishment, "Did you think this up all by yourself?" "Well, by golly, yes!" Although ranked only as technical assistant, I felt proud to be valued as a creative thinker by a guy who never doubted *his* future in science. Now thinking of *my* future, I tasted the pleasure of greater independence—designing experiments and interpreting results on my own.

While our results did not tell us just how nature contrives to generate one bona fide mating type from another, they did provide further insight into how they function.

It was only after the techniques of molecular genetics became applicable to *Schizophyllum* that all this information really paid off. To understand the nature of the genes involved, we had to clone and compare the wild type DNA with its mutated versions. When this was done—in *my* lab some two decades later—my two-gene-type concept gained confirmation.

Upon his return from sabbatical, Red decided to write a book summarizing everything we knew about *Schizophyllum* and related species. He wanted to call it *Sexes by the Thousands,* but the stuffy publisher deemed that too frivolous. They settled on the sober title, *Genetics of Sexuality in Higher Fungi.* The treatise has since been referred to as the bible of mushroom genetics.

I admired Red's disciplined approach, writing several hours every day in his office at Harvard and in the evening at home. I liked his writing, and wondered if I could ever succeed in such an endeavor.

Chapter Twenty-three

Faculty Wife

BEYOND THE RESEARCH, I rather enjoyed the privilege of being a faculty wife. It gave me the opportunity to mix with departmental members and their spouses, a fascinating lot with an array of interests and life experiences. My training as a sorority sister back at Syracuse had imparted skills of social mingling that served me well at large faculty gatherings such as the annual Prather Tea Reception for distinguished outside lecturers. Faculty wives hosted these pretentious affairs at the Harvard Faculty Club—now with its front door open to women. Canapés on silver platters decorated linen-clad tables while ladies of seniority, wearing white gloves, dispensed tea and coffee from elaborate silver urns.

Upon appointment to the Tea Committee, I asked: "Why not serve sherry, or even cocktails? Such a move might boost attendance and enhance conviviality!" The ladies agreed; it was time for change. Whereupon these Receptions increased in popularity—and often resulted in faculty groups sharing a jolly meal together while metabolizing alcohol before the drive home.

Despite the lubricating booze, a tendency remained for females to congregate in one group of clusters and males in another. Female talk covered a range of family troubles, social ills, artistic endeavors, and exotic travel. From one year to the next, the female exchanges resembled a kind of soap opera with yearlong interludes. The women seemed to relish this opportunity to socialize amongst themselves, away from their husbands.

Male talk centered on scientific breakthroughs, the latest jokes, and exotic travel, often delivered in lecture mode discouraging conversational exchange. I mingled, intermittently sampling both, sharing news and thoughts with the females but mostly just listening to the males. Clearly those high-powered professors thought well of themselves. Most talked more than they listened. One in particular never managed to give me more than a nod while loquaciously addressing his male colleagues. His wife talked incessantly when not with him but with other wives.

Still, several welcomed my presence. For example, George Wald, the biochemist, who got the Nobel Prize for his and his wife's research on vision, pontificated robustly yet did respect my questions and comments. His trust in my abilities would be acknowledged later, after Red's death, when he asked me to plan and conduct the microbiology labs for his legendary course in biology for non-science majors.

I would come home from such gatherings, give my husband a big warm hug, and say, "I'm so glad you're not like *them*!" Red had a lot of ideas, but he didn't expound much in social circles.

While we women revealed some personal tidbits, I chose to share only one of two momentous decisions that Red and I made in 1968. The one we felt free to share was a decision to quit smoking those miserable cigarettes we'd been hooked on for decades. Red, the longer and heavier smoker, quit his habit a week before I did. He bore some physical symptoms—shortness of breath, hacking cough. It frightened him to realize a recent statistic: The probability of suffering an early death from cigarette smoking exceeds that of being killed as a daily commuter on Route 2 from Lexington to Cambridge.

Faculty wife and husband

Nonetheless, his 37-year smoking habit probably hastened heart failure in the end. I noticed no ill effects in myself at the time; yet, as was discovered much later, I did incur some long-term lung damage.

My motivation to quit stemmed primarily from our children's disdainful attitude, expressed most profusely when refusing a request to fetch a pack of Camels from our stash in the kitchen cupboard. As one ex-smoking colleague proclaimed, "We ask our kids to toe the line, do their homework, stay away from drugs. The least we can do is set an example and muster the discipline to stop smoking those lethal fags." It took years to stop yearning for cigarette breaks with coffee, tea, martinis, and the like, but somehow we managed without relapse.

The other decision—not up for discussion with faculty wives or anyone else beyond intimacy—involved an unplanned pregnancy. Despite having taken the usual precautions for birth control then recommended by our family doctor, I became pregnant at 43 when our children were in their teenage years. Immersed in research at the lab, I did not want another child. Nor did Red. Our budget could not handle it, and any career in science for me would slip away.

I beseeched our doctor. "I did as you said. I kept track of my periods and always used the birth control cream you prescribed. What can I do?" He smiled, cocked his head, and rejoined, "No method of birth control is foolproof. You're a healthy woman. You'll have that baby and be just fine." *That's your opinion,* I thought, *you're not me. You're a career doctor with a stay-at-home wife. I don't want to follow my mother's path, spend the peak of my energetic years raising a lot of children the world doesn't need. I want to do other things, use my skills to discover new ways of understanding our world. You just don't understand.*

Of course I could not expect *him* to terminate my pregnancy—abortion being illegal then—but he might have offered some information as to where such a procedure could be done legally; he might have at least expressed some empathy. Though not of militant manner like the Chicago doctor who presided over our firstborn, this man embraced the usual 1960s concept of female roles. I thought back to a remark he had made about his belated discovery of our young daughter's slightly crossed eyes resulting in monocular vision: "Well, she's a girl; she won't be piloting airplanes."

Red and I considered the possibility of my traveling to some other country, like Finland or Japan, where abortions were then legal, but such long-distance travel seemed out of the question. We hadn't the money nor could I bear the thought of leaving the safety and comfort of home. I opted instead for illegal abortion nearby—a scary process involving intrigue from beginning to end.

I phoned my old college buddy Conant, thinking she might know of a number to call. Worldlier than I, she put me in touch with a Boston doctor who would perform the procedure anonymously in some undisclosed location, for a price. We agreed upon a date and time.

I was to meet an envoy—her face obscured by large dark glasses and a pulled-down hat, color green. She would arrive under the clock at Harvard Square. Red took me there. Standing in front of the Coop, in the rain, we watched and waited together. The envoy came; I kissed Red's cheek, squeezed his hand, and joined her. She handed me a pair of wrap-around opaque glasses to cover my eyes. Taking my elbow, she escorted me like a blind invalid to a nearby car that whisked us away to some place in the city I could not discern.

The car stopped. She grasped my arm and took me across what must have been a sidewalk, then up stairs to the inner sanctum of what I imagined to be the second floor of an old brownstone building. After lowering me down to a straight-backed chair, she removed the glasses that had served as my blindfold, and money changed hands—more than it might cost to deliver a full-term offspring. She retreated through a rear door, leaving me alone.

My seeing eyes adapted to a dingy-looking sitting room with dull green walls. A hospital gown lay in a chair next to mine. I changed and waited in silence, feeling nervous and humiliated. Soon, a masked doctor appeared, took me into his operating room (probably a converted living room), put me on a sheet-clad table befitted with stirrups, and performed the deed. He seemed competent enough. His eyes conveyed age and experience. He commented afterwards, "Yes, there it is, about a six-week-old embryo." Upon release, I got dressed, was blindfolded by the glasses again, and was escorted back to Harvard Square, where Red waited to take me home to bed and to a bowl of hot chicken soup.

Fortunately, no complications ensued, but what if they had? My story ended without disaster. I know of others whose did not. In light of ensuing events, I never regretted our choice.

Had our unbidden offspring come to term it would have been only six when Red died.

THOSE LAST FEW YEARS I had with Red marked good times. As the kids became more independent with their own activities, I spent more time at the lab, taking an increasingly independent role in the research. With Red's encouragement, I had gained a great deal of confidence in my own abilities to think and do. For diversion Red did his shop work; the kids did music and theatre; I sang with the Masterworks Chorale, and took the kids skiing. And our terrier bore a batch of entertaining puppies that all found good homes.

We all car-camped out west for two different summers to share the wonders of national parks, monuments, and forests. Then for the first time since the beginning of parenthood, Red and I left the kids on their own while we sampled a Caribbean isle or two—the only time we'd gone abroad for non-scientific reasons.

The two remaining years with Jonathan and Linda away at college were like a second honeymoon.

Second honeymoon

As much as I loved our kids, I suffered not the empty nest syndrome. They were off and growing up; their parents no longer needed to quarrel over disciplinary requirements. My experienced mom had a saying about raising children: "By the time they're sixteen you've done all the damage you can; you might as well give up."

But then, after Red's death, I mourned that empty nest terribly.

Chapter Twenty-four

My Own Lab

NOW RED WAS GONE. Upon my return from the trip to Europe—not with him as planned but with daughter Linda—I settled into Red's office and lab space overlooking the twin rhinos bracketing the Bio Lab's front doors. Upon application, a grant from the Campbell's Soup Company to study the sex life of the common edible mushroom *Agaricus bisporus* and related species had been transferred to me, thus allowing continuation of research Red and I had started four years earlier.

We had moved from work on sex hormones in the water mold *Achlya* to a series of investigations on mating and sex in members of the mushroom-bearing fungi—first on the amenable wood-rotting mushroom with multiple sexes, *Schizophyllum commune*, and more recently on the intractable crop mushroom, *Agaricus bisporus.*

A. bisporus is a fussy fungus; it does its business slowly in a rather cryptic manner. Nonetheless, we, in collaboration with Bob Miller, a scientist at the Campbell Research Institute, did expose its sexuality: homosexual. This popular palatable mushroom had been cultivated since the 18th century—King Louis XVI purportedly kept a patch in his palace garden—yet no one before our study had found clear

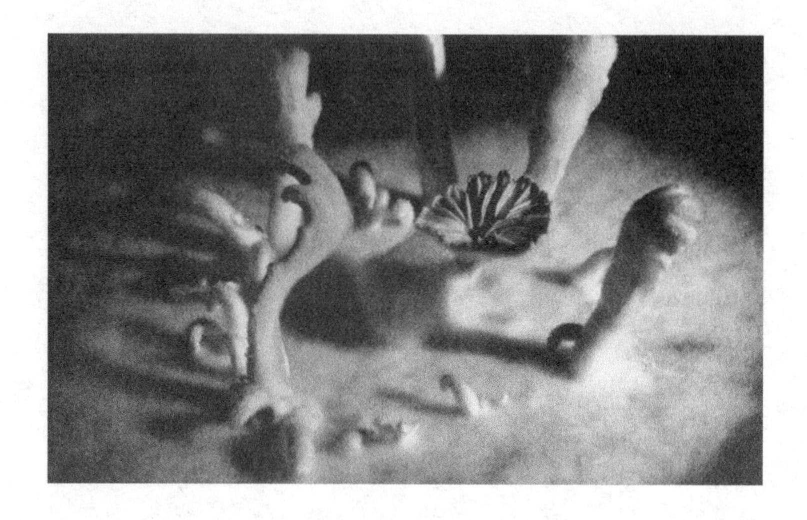

S. commune *in cultivation. Published in "Genetics," Vol. 129, by Horton, J. S. & Raper, C. A., titled "A Mushroom-inducing DNA Sequence isolated from the Basidiomycete, Schizophyllum commune," pgs. 707-716, Copyright Elsevier (1991).*

Harvesting the button mushroom A. bisporus *from compost*

evidence of its ability to mix genes through a sexual cycle. Throughout nearly three centuries, strain improvement relied upon the inefficient process of selecting randomly occurring spontaneous mutants that enhanced mushroom production.

We thought a genetic approach might reveal a clandestine sex life useful to breeding superior commercial strains.

Transfer of the Campbell Soup grant to me afforded support of two invaluable associates: Margit Bensen and Gerry Kaye.

Margit had been one of the most essential members of Red's lab for over 10 years. Maker of media, custodian of the glassware, she also played an essential role as resident psychiatrist. Without Margit the whole place might have fallen apart. She brought her wisdom and skills to us, having emigrated from Norway some 35 years earlier. Aiming at first to upgrade from midwife in far northern Norway to registered nurse in the USA, she had not realized the importance of mastering the English language beforehand.

Margit never did become a medical nurse; she nursed our lab instead. It was our good fortune to have this uncommonly wise, witty, giving woman join us after a series of other positions—assistant cook in a Swedish restaurant, assembler of corsets in a corset factory, and upstairs maid in a U.S. senator's home on Beacon Hill.

Margit came to us as a part-time tech while also working part-time in Jim Watson's lab (the DNA double-helix guy) just down the hall. Realizing her value to our group, Red found the funds to hire Margit full-time. While Margit had a fondness for Jim, she loved Red like a mother and decided to work for us exclusively. She had her own kitchen with autoclave, dishwasher, and sink, over which hung a little wooden puppet she called Yimmy Boy. She emphasized amusing anecdotes about Yimmy Boy by yanking on the puppet's dangling string to make it flail its arms and legs as Jim might do—a mocking gesture not meant to be unkind.

Margit presided over her domain as a mother might in her own kitchen. Generally, students and techs found themselves at that awkward stage in life sans parents, yet still wanting a confidant for troubling thoughts. Margit, the perfect listener, engaged each needy student in helping with the washing up, fitting bottle tops, stacking glassware, and loading the lab washing machine—a process not requiring eye contact as personal problems got discussed.

The young folks did most of the talking while Margit did most of the listening, occasionally interjecting empathetic comments spiced with charming good humor. They knew she would keep their secrets. With soft red hair, a kindly face and huggable figure, Margit reached beyond the lab. She often invited us in groups to her home in South Boston to meet her husband Arne, and enjoy her special Norwegian dishes like lutefisk and krumkake. She had no children of her own—we from the lab became her children, our children her grandchildren.

Margit

Gerry Kaye had been a tech in the Raper lab for more than a year before Red died. We hit it off the moment she arrived. Smart as a whip, she loved fungi, had good hands at the bench, and superb secretarial skills. As mother of two children at the age of relative independence, Gerry happily emerged from full-time homemaking to something more intellectually challenging, a concept beginning to take hold in the 1970s.

Ours was the only all-female lab on the premises. We had the money. We had the space. We had the expertise. We could *do* it!

Although Red, Bob, and I had previously discovered sexuality in the button mushroom (now affectionately dubbed *Agaricus groceri*) we realized the difficulty of breeding this self-fertilizing species to make bigger, better, and healthier mushrooms. Accordingly, we wondered whether or not some related species might taste just as good but be easier to handle. By the time of Red's death, this investigation was well underway—with me taking the lead.

Fungi exhibit enormous diversity in lifestyles, including their sexuality. Some are haploid, with a single set of genes; some are diploid, with a double set of genes; and those with either type of genetic makeup may be either sterile, possessing no sexual cycle, or sexy. Among the sexy fungi, some can fertilize themselves while others require a mate; some can even do both.

Agaricus groceri fertilizes itself exclusively, as would a hermaphroditic animal but without specialized sex organs. The only clear morphological manifestation of this particular lifestyle is the formation of succulent mushrooms. But self-fertilization preempts mating with other individuals of diverse and possibly more promising potential. A heterosexual ("straight") species requiring a mate for fertile interaction would be the way to go. We thus began examining other species of *Agaricus* and comparing their morphological and physiological characteristics, with an emphasis on succulence and sex.

Morphology ruled taxonomy within the genus at the time. Some of the defining visible characteristics such as size and shape of spore, color of cap, and condition of fruiting, could be due to a single gene difference. We thought other criteria should be considered, especially ability to interbreed. Our work to date had identified one promising species, *Agaricus bitorquis.* Cultivatable on a commercial scale, it produces tasty mushrooms and is easily bred. Our all-women team now continued that work and extended the study to other related strains collected from nature.

Harvard's generosity in allowing us to occupy so much of Red's old space with a view of the rhinos did not last. We had to make room for a newcomer with a larger group. Margit, Gerry, an undergraduate Peter Haven, who had joined our lab, and I moved one floor down to smaller but adequate quarters looking north to a parking lot.

Then came the honor of my appointment as Lecturer to design and prepare a series of labs for a legendary course called Nat Sci V (Natural Science) meant for non-science majors. The lectures, delivered with a flourish by George Wald, attracted large numbers of students required to take at least one science course. According to custom, seasoned full Professors taught not only graduate students but undergraduates as well. Their undergraduate teaching, however, seldom extended to the laboratory part of the course. Junior faculty or the rare so-called Lecturer such as myself, planned the labs and instructed teaching assistants drawn from the ranks of graduate students. Wald asked me to design several lab sessions relevant to microbiology.

Recalling my own experiences as an undergraduate student of science, I decided to avoid the usual cookbook-type lab exercises and prepare sessions in which the students must puzzle out an unfamiliar problem. I wanted them to experience a sense of discovery rather than simply follow prescribed recipes to obtain previously produced results.

Red had shown me the way with his hands-on approach. I was influenced also by exposure to a world-renowned teacher of microbiology, Cees van Niel, under whose tutelage Red had spent one summer back in the early 1950s.

Red's switch from water-mold sex to mushroom sex had made enough progress by the summer of 1951 (two years after our marriage) that he thought it time to learn more about teaching. With a quarter free from course work in Chicago he aimed to hone his teaching skills by learning from a master. Stanford Professor Cees B. van Niel, the master of choice, taught a famous six-week course in microbial biochemistry each summer at the Hopkins Marine Station in Pacific Grove, California. He limited the course to about a dozen students at the graduate or early professorial level. Its popularity ensured an excess of applicants.

Privileged to be accepted, Red headed cross-country in our year-old elephant-colored Chevy on the first of July. A hitchhiking student and our recently acquired wire-haired fox terrier, Soo Pooh (Soozie), rode along with him. Not having the whole summer off from my work at Argonne, I stayed home until mid-August, when, it was planned, I would join him out west.

Red wrote nearly every day. Despite his having our dog for companionship, he missed his woman; I missed my man *and* the dog. Red wrote:

> I had had rather serious doubts coming across Utah and Nevada, about the worthwhileness of this venture compared to what I could have been accomplishing back in Chicago this summer. After one day's session in the course, however, I have little doubt that it will mean a lot to me when I initiate something at a lower plane at Chicago next winter. Van Niel is an out-and-out ham, tremendously impressed with his [own] cuteness and histrionics. But even on short exposure there's no question about two things—and for these I can forgive him a lot: He's sincerely interested in his students and his subject, and he's goddamned effective in getting the stuff across to them in a hell of an impressive and digestible manner.

I think when you come out, that you might very enjoyably and profitably sit in on a few sessions.

Mid-August finally arrived when I could leave my job at Argonne and join Red on the Coast. We had counted the days. The question arose, should I fly and get to my love in a hurry, or sacrifice another day's absence by taking the rail route on the Burlington Zephyr? With Red's acquiescence, the latter prevailed.

Leaving Chicago in late afternoon, the Zephyr sped by night through Midwest flatness to reach Denver at dawn and get its windows scoured for clean views of the majestic Rockies. After snaking through the Great Divide in the brilliance of daytime, our train moved westward through vast moonlit deserts that second night. I stayed up late dreaming of coming romance as I gazed half-dazed at the passing landscape. I felt quite the sophisticated lady, traveling such a distance all by myself.

Red met me next day in the dry heat of Sacramento. I'd changed to high heels and a form-fitting, black, sleeveless dress in preparation for our long-awaited reunion. We hugged and kissed our way to the car, which to my surprise had been all closed up with the heater in readiness. As Red explained, we'd soon be over the coastal range in deep fog, where heat would be welcomed.

We settled in at the charming, bougainvillea-covered cottage Red rented from a characterful landlady named Kippie Stewart in Carmel-by-the-Sea. It was just as he had described, a seven-foot bed, wicker chairs, and a fireplace set for a roaring fire. The slinky black dress somehow got ripped that afternoon. When I asked for a needle and thread in hopes of repairing it, Kippie cheerfully volunteered the mending herself.

In the remaining week of Cees van Niel's course I saw for myself its reason for fame. Van Niel mentored as none I had known before. He literally threw himself into that course. He spent as much time preparing for it as he did in teaching it. I was told he collapsed exhausted and went to bed for a month afterwards to recuperate.

A typical session started at 8 AM with van Niel lecturing energetically, posing problems to be solved while cleverly coaxing his students to seek solutions through experimental means. In one session, for example, he presented the case: "A walk along an ocean beach at night sometimes reveals luminescence where one's feet kick up the sand. Now what do you suppose might be the cause?" This opening gambit put students on alert. They volunteered a variety of answers, such as moonlight reflected by small chips of pyrite or agitated bits of flint setting off small sparks. Each was considered and then dismissed as being rather doubtful.

Discussion ended with the suggestion that some microorganism capable of producing light might live in the sand. "Well, yes, that is a plausible theory," allowed van Niel. "Now, how might we test it?" Again, he skillfully elicited various ideas from members of the class until someone suggested going to the beach at night, checking it out for the presence of luminescent sand, then taking a sample back to the lab for testing.

"But how would you test?"

"We could spread the sample over the surface of a variety of nutrient media and see if anything grew. Any growing colonies could be checked for luminescence in a dark room and isolated to fresh media."

"But what if something grows and emits no light? Then what would you do? Think about it; do we see any light coming from undisturbed sand?"

The students realized, the answer is no.

"Now what? Maybe some of the growing colonies are capable of luminescing but can do so only when agitated. How can that be tested?"

After further group discussion, someone came up with the idea that each different colony of bacteria could be inoculated to a test tube containing a liquid version of the appropriate medium and, after incubation, be observed for evidence of luminescence while being agitated in the darkroom.

Whereupon, van Niel, nodding approval, disappeared into an adjacent room and retrieved a lab cart previously loaded with all the necessary equipment for the students to perform the "spontaneously" suggested experiments without delay.

This performance illustrated van Niel's utter confidence in his ability to lead the class to arrive at a certain conclusion, all the while teaching not only facts but methods of achieving those facts. His next session included a lecture presenting evidence about how the chemical responsible for the emission of light in luminescent bacteria is dependent upon oxidative processes of respiration that usually increase upon agitation.

I could not hope to emulate Cees van Niel, but I wanted to catch some of his flavor in the labs I planned to design.

Chapter Twenty-five

Sauerkraut, Mushrooms, and Penicillin

As MY VERY FIRST teaching experience, I wanted my labs for the Nat Sci V course to be outstanding, to be unique in the manner of Red Raper and Cees van Niel. I worked diligently, hoping to justify Wald's confidence in me—perhaps overreaching.

My plan had three parts. The first and simplest started with shredded cabbage and ended with sauerkraut. Each student made their own batch and observed the changes in numbers and kinds of microorganisms coincident with the fermenting process as it evolved from sweet to sour over three or four weeks. The product could be savored at the end if desired.

It seems simplistic, and it was; but none of these undergraduate non-science majors had seen all those squiggly microscopic critters before, nor had they known their effect on something edible. It provided an opportunity to teach something about the process of fermentation used not only for sauerkraut, but yogurt, beer, and wine as well.

A second project presented the students with single-spore isolates representing different mating types of a couple of unnamed mystery fungi: two compatible strains of *Schizophyllum commune* with four mating types versus two compatible strains of *Polyporous betulinus* with only two mating types. We had them make matings in all possible paired combinations. They were to note any morphological changes as each pair interacted over a week's period, then try to make some sense of the patterns observed.

This exercise perpetrated puzzled discussion, leading to the enlightenment of not all students. Alas, few possessed the acumen of Herr Doktor Professor Hans Kniep, the initial discoverer of such sexual patterns. Though most students were undoubtedly interested in human sex, few mustered much enthusiasm over fungal sex. This project was a mistake.

The third idea met with greater success. We gave each person an opportunity to isolate a strain of *Penicillium* from the environment and test its ability to produce the antibiotic penicillin in the lab. Red had shown me how to trap fungi from the air and make a "mold garden" from which individual specimens might be isolated and examined.

"A vast variety of colorful fungi can be caught on a moistened slice of homemade bread which, unlike store-bought bread, contains no noxious chemicals designed to discourage the growth of any self-respecting fungus," said Red. "The spores of molds lurk everywhere," he continued, "kitchen, bathroom, bedroom, basement. Once settled on a bit of moist nutritious substrate, they go to town, germinating and developing into colonies of hyphae that converge to form a veritable garden of orange, green, black, and white patches."

Never having made bread, I modified Red's methods by substituting a semisolid agar medium made fresh by the lab techs. Accordingly, each student received a petri dish containing nutrient medium and was told to remove the lid, expose the dish to a place of his or her choosing for a day or two, replace the lid, and allow development of mycelial growth until the next lab session.

Mold garden

After examining the mix of colonies in class, we showed how the students could identify and isolate *Penicillium*. They tested the isolated colonies for their antibiotic potential.

This is where I called upon the expertise of Red's older brother Kenneth, who, during World War II, had been among the first to facilitate production of penicillin in commercially acceptable quantities. Ken provided Fleming's original strain—Fleming discovered it as a contaminant to a bacterium he was studying back in 1928—to compare with his own improved strain derived from a moldy melon during World War II. The strain was found in the marketplace by his assistant, thereafter called "Moldy Mary."

Of the four lab sections of the class of about 100 students, three came up with strains that produced penicillin in greater quantities than Fleming's strain. None surpassed

Mary checking moldy food for Penicillium

Ken, on left, and colleague discovering a superior penicillin-producing strain

Kenneth's strain, which was then being used for commercial production.

All together, this unorthodox approach to hands-on laboratory exercises excited me but failed to stimulate adventurous attitudes in a majority of the non-science majors. It was my first experience in trying to convey a passion for the ways of science to those who had never been enamored of science in the first place. My written evaluations from these undergrads indicated a distressing preference for being told precisely what to do in lab, how to do it, and what to expect in the end. Most seemed baffled by my attempts to stimulate independent thinking.

This first experience in undergraduate teaching taught me humility: I could not henceforth assume that what appealed to me would necessarily appeal to others. Nonetheless, one of the graduate student assistants, Jay Dunlap, who helped teach the labs, admired my approach. He remained a lifelong friend and colleague and later, in collaboration with his wife Jennifer Laros, pioneered the discovery of circadian rhythm (e.g., reasons for jet lag in humans) in a fungus called *Neurospora.*

DURING THAT FIRST YEAR after Red's death, the women in the Bio Labs began weekly brown-bag lunches to discuss various problems pertaining to being female in a male-dominated profession. The group was open to all female employees, whether stockroom keepers, technicians, students, or faculty. It included the sole tenured female in the Department, George Wald's wife Ruth Hubbard, and four or five female junior faculty. We munched our lunches sitting in circular fashion to encourage ambiance.

Discussion ranged freely from sexual harassment in the elevators to patronizing comments from Departmental males. Complaints of the former sort seemed so common as to make me question my degree of attractiveness for not having experienced such treatment. I thought up excuses: could my status as the late Chairman's wife explain a reluctance to flirt?

Awareness of the concerns of other females in the Department, however, led me to think seriously about long-range plans for developing a career as independent scientist without the benefit of Red's mentoring and protection. Our children no longer needed much parenting. But with Linda in college and Jonathan in graduate school they did need financial support, some of which I had to supply.

Chapter Twenty-six

PhD at 52

AFTER RED'S DEATH I felt the strong need to continue our work with fungi. First attempts to fulfill that need exposed my naiveté. Knowing I would have to achieve a more permanent status, I investigated the possibility of staying at Harvard as Research Associate and keeping the lab going by obtaining continuing funds from granting agencies. The realization that a Research Associate could not function as an independent biologist within the Faculty of Arts and Sciences promptly squelched this idea. Only the Medical School permitted such privileges.

I next tried to determine whether or not I might affiliate with some tenured or tenure-track professor. Well, no one wanted to give up space and jeopardize his own chances of funding by sponsoring someone else in an unrelated project. What then *could* I do?

Lawry Bogorad, a longtime friend from my grad student days at the University of Chicago, who had since networked his way up from student and Assistant Professor of Botany at Chicago to Full Professor at Harvard, helped me see the light. When asked about prospects for continuing my work at the Harvard Bio Labs, Lawry put on a smile, patted the top of my head, and proclaimed, "Cardy, you don't want to enter Harvard through the back door. You'd always be considered second class. You need to get the usual credentials and look for a job just like the rest of us."

I felt belittled at the time, but Lawry made sense. His was a timely suggestion. Others earlier, including Larry Powers, wondered why I didn't go on for the doctorate; Red and I discussed it. But to this point I had not wanted to interrupt the work I enjoyed so much; even now, I think of those 12 years in Red's lab as among the best. I loved doing the research without the responsibility of having to keep the whole enterprise going. Now without Red's support, I realized the necessity of obtaining a PhD.

Two influential females helped me steer the course: Ruth Hubbard (George Wald's wife and the only fully tenured woman in the Department) and Ursula Goodenough, an Assistant Professor full of vim, vigor, and more talent than I could ever hope to have. Both women agreed I had to get a PhD or what they referred to as "the union card." Ursula, whose field of research on the genetics of mating in the alga *Chlamydomonas reinhardi* related to mine, said she'd be honored to sponsor me in my work on *Agaricus* species.

With trepidation, I approached John Dowling, chair of the Graduate Committee in the summer of 1975. He seemed to know my purpose and put me at ease. "Certainly," he said, "the Committee will consider your request to enter our graduate program in biology."

Lo and behold, the Committee approved my application, and waived tuition to boot. Furthermore, its members did not prescribe the usual exams for determining what courses should be taken in preparation for the final qualifying exam and thesis defense. They thought me mature enough to know how to obtain the knowledge I needed; they gave me permission to do this at my own discretion, whether through independent study, auditing, or taking courses for credit. Their trust in my abilities exceeded my own, but I was gung ho to give it a try. Positive forces had shifted to first gear!

Thus began a year in which hard work left little time for play and sleep. According to the Greek definition of philosophy, a major part of my life would be devoted to "the acquisition of wisdom through a process of rational inquisition." The Department's confidence in me put to the test my strong desire to continue Red's work as his legacy. Yet I questioned my own ability to carry on without him. Rather than succumb to such self-defeating thoughts, I *had* to keep trying.

With so much to learn in so little time, I feared my brain resembled a leaking sponge. I had prided myself in recognizing themes of Beethoven quartets, names of Supreme Court Justices, heights of Adirondack peaks. Now, that knowledge receded, buried under the verities of molecular biology: DNA to RNA to proteins, and known activities in between. I had to work so very hard!

The details of everything that happened during that year of graduate study have gone from my memory. In broad outline I recall a course in biochemistry brilliantly taught by Mark Ptashne and Guido Guidotti; a course in cellular biology also taught brilliantly by Dan Branton and Win Briggs; a scramble to keep my assistants Gerry and Margit busy with the research; lively luncheon sessions as part of Ursula's lab group; and regular calls to son Jonathan in which we commiserated about the trials and tribulations of graduate-student life.

Daughter Linda quit Bennington College at the end of the fall semester in the year after Red's death. She had chosen to go there because of its strong program in the arts but more significantly because Bennington was the only college to accept both her and her high-school sweetheart Leslie (of the Lexington Green night)—they simply could *not* be separated. They separated within the first year.

Lin felt the need to get on with her life. She needed a purpose. She loved small children, and decided to transfer to Wheelock College in Boston, where she could graduate with a degree in Early Childhood Education. She moved in with me. I wanted her presence. We truly needed each other.

By the fall of 1976, just a year after I entered graduate school, it appeared that I'd be able to complete my thesis, face the oral qualifying exam, and finish my doctoral training at the end of January.

The thesis, entitled "Comparative Biology and Sexuality in *Agaricus*" consisted of four publications with written introduction and discussion. One, jointly authored by me, Bob Miller, and Red, centered on the crop mushroom *Agaricus bisporus.* Two others, including a study of the more breedable *Agaricus bitorquis,* were solely mine. Gerry and I coauthored the fourth. It comprised a comparative study of 15 morphological, sexual, physiological, and chemical characteristics of 33 specimens of *Agaricus* collected from nature. It took weeks of tedium to work out how to correlate all these data without the benefit of computers, but we did reveal some hitherto unknown taxonomic relationships.

Since each publication had its own introduction and discussion section, I deemed it unnecessary to add much. Ursula deemed otherwise, and sent me back to the drawing board for a more comprehensive draft she might agree to accept. As I look back to the fits and starts of producing this thesis without benefit of computer technology, I marvel at the speed of its accomplishment. Without Gerry Kaye's expertise as intelligent technician *and* typist, it would no doubt have taken much longer. I benefited enormously from Ursula's disciplined encouragement. She, some 18 years my junior, became a role model I never could match. To date, she's had three husbands, five children, a popular genetics textbook, a best-selling book on science and religion, and hundreds of scientific publications. Well, all I could do was try with what I had.

One dark December afternoon, the five members of my Graduate Committee convened in the dreaded "Gas Chamber" to administer a final test. I knew they were friendly and wanted my success; nonetheless, the whole process held an element of terror. Among the watchful professorial portraits lining the wall hung one of my beloved Red, looking

Portrait of Red.
(Printed with permission of the Department of Molecular and Cellular Biology, Harvard University.)

at me rather critically but with a twinkle in his eye that said, "You can do it, kiddo!" This was the defining moment: sink or swim.

I had no trouble speaking, but the Committee's queries demanded a certitude of knowledge not always recoverable. I remember only one question for which I had no inkling of an answer: "How do you determine the ionic strength of a solution of potassium chloride?" "I don't know." The questioner provided a hint. I still could not answer. He kindly let it pass, noting that physical chemistry was neither my major nor minor.

Afterward, I paced the hallway waiting for a verdict. Ursula emerged within ten minutes and welcomed me to the club: "Congratulations, Doctor Cardy!" We celebrated with a rip-roaring party, and, unlike newly minted Dutch Doctors, I did not have to throw it myself.

Having previously arranged for a postdoctoral year with Professor Jos Wessels at the University of Groningen in northeastern Holland starting in February 1977, I was ready to move on to a new level.

Chapter Twenty-seven

Dutch Postdoc

My doctoral work completed, Linda and I chose to celebrate my last evening at home by going out for dinner. We'd reserved a table at a favorite restaurant called Henry IV in Harvard Square. As we walked from the warmth of our car to the warmth of the restaurant, a winter wonderland built around us with snowflakes so luscious we tilted our heads to catch them on outstretched tongues. The ambiance of that restaurant with its candlelit, linen-draped tables set a festive mood. Linda and I never lacked for animated conversation. Excitement in anticipation of my impending journey cast a pleasant aura over a savory meal of *pot au feu*. We knew, nonetheless, we would miss each other terribly during the ensuing months.

My journey abroad started with a side trip to Israel. Red's former grad students Yigal Koltin and Judy Stamberg had invited me to give a seminar at the University of Tel Aviv, all expenses paid by the U.S.–Israeli Bi-national Grant, to which I was named as the United States counterpart after Red's death.

In direct contrast to the snowy streets of Cambridge, Israel's sunny clime enhanced my transition to a different culture. True to form, Yigal and Judy ushered me about from dawn to midnight, jamming in as much shoptalk, sightseeing, and partying as possible. My seminar elicited lively discussion that continued as I walked with students about the campus, talking about their research and mine. For the first time I found myself viewed as a mentor and principal investigator apart from my husband.

After a day's visit north to the Sea of Galilee, Yigal and his wife Sara held a gathering of colleagues in their home for a party in my honor. Although scheduled for 8:00 PM, no one arrived before 9:30. Explanations ranged from "The children had to be put to bed" to "We had to finish watching a TV show that started at 8:00." Such cavalier attitudes bore little consequence, since tasty snacks appeared for late-night consumption and everyone stayed to profess strident opinions on science and politics through the wee hours of the morning.

Exhilarated but not feeling rested from several days of visiting with Yigal and Judy, I went on to Jerusalem at the invitation of two other colleagues, Yair Parag and Giora Simchen, both of whom had spent time in our lab at Harvard and had later hosted Red on his last sabbatical leave in Israel. After my seminar there I managed to rest a bit at the Simchens' home, since they, unlike Judy, Yigal, and Yair, embraced a calmer, less agitated view of their country and its precarious position in affairs of the world.

At the Simchens' suggestion I spent two days on my own exploring the ancient city of Jerusalem, an experience so absorbing as to occupy all senses all the time. On the second day, sights, sounds, taste, and smells became so all-consuming that I tripped on a bump in the street and sprained an ankle. The wounded ankle had a chance to rest the next day when Yair drove me to the Biblical site of Masada overlooking the salty Dead Sea, all the while imparting his vast knowledge about the history of the region and the zealots who defended that lofty fort to the death, decades before the time of Christ.

Israel is so rich in old and new ideas that a quick tour seemed hardly sufficient. I departed with strong feelings of a nation on edge, ever vigilant, ever fearful of its continuing existence, ready to protect itself as zealously as the zealots of Masada. For me, the renewal of friendship with Israeli colleagues promoted a lifelong sense of mutual affection, based now on a more equal footing as scientists with common interests.

Upon arrival in Haren, a pristine village in the picturesque northeastern part of the Netherlands, I was shown my future quarters—a light-filled bedroom with kitchen privileges in the home of a cheerful chatty widow and within walking distance of Professor Wessels' lab at the University of Groningen. It would be my first experience alone away from home for an extended period of time. Sans automobile, I depended upon feet, bicycle, busses, and trains to go anywhere.

I soon discovered a disparity in the way Dutch scientists function compared to Americans and Israelis. Contrary to the long, often irregular hours I spent in the lab and office back home, Dutch scientists seemed to live less hectically, perhaps more efficiently, working on an 8:00 to 5:00 basis five days a week, with time out for lunch and recreation. They left promptly at 5:00, carrying briefcases serving more as props, I thought, than working entities.

Old habits did not die. I worried about safety while working weekends alone in a large lab complex within the deserted building encompassing Jos Wessels' Department of Developmental Plant Biology. What if I slipped and knocked my noggin on a bench top or autoclave door? No one would find me before Monday morning.

We worked out a plan: Jos agreed that I should phone him

upon arrival at the lab on a Saturday or Sunday and say how long I expected to be there, then call again as I left. Should I fail to call at my estimated time of departure, he would phone or come by to check. That worked, although I never did bump my noggin on anything.

Weekends tended to loneliness until I'd established some outside activities such as singing in choral groups, taking an art class in drawing, going on sightseeing trips by train, and making music on my fipple flute. I soon found a lifelong friend in Tilde Tinbergen-Frensdorf, a Jewish refugee from Germany in World War II, and the widow of a famous Dutch scientist. We played our recorders together in ensembles and enjoyed long bike rides over the countryside.

I learned to take daughter Linda's advice: "Mom, don't be afraid to treat yourself now and then. Take yourself to a nice restaurant; go to museums; explore the country." Although living alone in another's home harkened back to college-student days, I began to enjoy the advantages of being on my own. Brother Lute, who had always lived by himself, reinforced Linda's advice. He served as a role model during this year of independent living. He pretty much did as he pleased and enjoyed it, going to concerts, reading, meeting friends, eating out. I thought back to the time when Red was in Israel, and the kids and I could indulge our moods without the need of accommodating his. Yes, there are advantages to living alone, not beholden to anything or anyone but work.

Abandoning *Agaricus,* I focused once again on *Schizophyllum.* The Wessels' lab had recently developed the long-sought technique of rendering *Schizophyllum*'s hyphal fragments into protoplasts, little balloon-like entities devoid of cell walls that could be manipulated as single cells and, under the right conditions, regenerated into multicellular hyphae with walls. We knew then that the generation of viable protoplasts was a necessary first step to gene cloning, because the fungal cell wall is impervious to large molecules such as DNA. But several years' work would be required before cloning became a possibility.

My postdoctoral year started with learning the complex business of producing, manipulating, and regenerating protoplasts. Having accomplished that, I wanted to try and fuse them to investigate the influence of cytoplasmic (nonnuclear) elements on sexual development. The experimental plan involved fusion of protoplasts with and without nuclei. Unlike membrane-bound animal cells, the membrane-bound fungal cells refused to merge. I couldn't even get past the control experiments. Despite this failure, the seven months in Wessels' lab left me with a wealth of expertise I would use to great advantage in the years to come.

Herr Doktor Professor Karl Esser, the mycologist with whom Red had done his first sabbatical, invited me to visit him and give a seminar at Bochum University while I was in the Netherlands. Now, as head of the Botany Department and curator of the University's botanical gardens, he proudly showed off his entire domain. I was pleased and flattered by so much attention until he patronizingly patted my hand over dinner and offered to *help* me write and publish my current research. "Now, on your own, you must publish, you know. As chief editor of my scientific journal, I can assist you; just send me a draft and I'd be glad to go over it, put it in publishable shape . . . "—the implication being that I must be incapable of managing this by myself. Karl probably meant well, being born and bred in a male-dominated country where Departmental Heads in academia were virtual lords of the manor, yet I took offense. His attitude spurred me on to seek a secure position as independent head of my own lab back home.

Toward the end of my stay in the Netherlands, I grew lonesome for family. Daughter Linda graduated from Wheelock College in June of that year, and I could not wait until August to see her. My beloved mother, who cherished *her* only daughter, had died the previous fall and left a small inheritance to each of her six children. I thought it fitting to use some of that legacy for *my* daughter's graduation present: a summer in Europe backpacking from one youth hostel to the next.

I doubt I could muster the courage for such travel myself, but Linda relished the thought of independence while mingling with others from other places. I worried about her safety, especially in Italy with her fair skin and long blond hair. She nonetheless managed an artful lightheartedness in fending off suitors in that cheerful clime. Linda handled an incident in the Paris Metro, however, somewhat less elegantly. After being followed by a seedy-looking guy through several transfers, she turned around and confronted him in a dark tunnel. She planted her feet and said, "Voulez vous un coup de pied dans les boules?", thinking it meant, "Would you like a kick in the balls?" The words, while inaccurate, served the purpose.

I was proud of Linda's ability to manage on her own. I think we were teaching each other lessons then, just as Jonathan and I had taught each other a thing or two as concurrent doctoral students.

The European adventure ended for both of us when we met in England at Oxford University to hear a performance by the Mastersingers on tour—a choral group to which we both had belonged back home in Lexington. We sat in the balcony of the performance hall holding hands in celebration of reunion, and cried—something that mothers and daughters can shamelessly do.

LINDA FLEW HOME TO Lexington while I went first to the University of Florida, Tampa to participate in a symposium dedicated to the memory of Red Raper at a meeting of the American Institute of Biological Sciences. It was like Old Home Week: A dozen of Red's former students and colleagues assembled; each delivered a research paper to be published in a small book entitled *Genetic and Morphogenetic Studies in Higher Basidiomycetes.* The book bore the inscription "To the memory of John Robert Raper, scientist, teacher, friend—not so much the words . . . but the ideas." The event called to mind all the times I had missed Red over the past seven months in Holland. It touched me deeply. I took pride in being considered an equal among experts in the field—and first speaker at that. *Yes, I can do it. I can go on without Red.*

Among others, Jos Wessels, Yigal, and Judy were there. Red's former Mexican grad student Celia Dubovoy also contributed. She gave a fine talk about relationships among a cluster of genes that block the process of fertilization in *Schizophyllum.* Recently recovered from a terrible burn accident she suffered as postdoc in Norm Giles' lab at the University of Georgia, she had moved back home to become Assistant Professor of Mycology at the University of Mexico and seemed more content than ever.

The happy memory of that reunion is sadly tainted by a tragic postlude. On her way back home, Celia boarded a plane that developed serious mechanical problems on the first part of her flight from Florida to Mexico City. As I learned later from her father, Celia had stayed calm and tried to comfort other panicky passengers in fear of crashing. The plane returned to base. She boarded another flight, arrived home, and went straight to bed—never to rise again. Celia suffered a fatal heart attack two days later. She died at the age of 29.

As I understand it, an autopsy revealed that her lifelong physical disabilities stemmed from heterozygosity for the genetic diseases Tay-Sachs *and* Gauchers. The two disease-causing alleles together, although normally recessive, somehow resulted in serious infirmities despite all the medical care she was given throughout her lifetime.

Whenever I start to pity myself because of some ailment or disappointment I think of Celia and how brave she was to do her work in the face of such difficulties. We came to realize her most important scientific legacy a decade later, when a student she had mentored characterized a mutant gene essential for the development of a method for gene cloning in our organism.

Chapter Twenty-eight

Job Hunting

BACK ON BASE AT the Harvard Bio Labs, my credentials completed, I had to find a job. My curriculum vitae listed 19 publications, including 17 peer-reviewed research papers (three with me as sole author) and two invited book chapters on the edible mushroom. It all looked pretty good for a first-time applicant to academia, but my age and gender were handicaps in the late 1970s.

I dutifully made a dozen copies of my CV and sent them off to institutions advertising openings conceivably related to my interests and qualifications. The CV revealed glaring discontinuities: bachelor's degree 1946, master's 1948, PhD 1978, first publication 1955, next publication 1964. The lacunae reflect times out for motherhood, civic responsibilities, musical pursuits, and trying to be a good wife. Would I be wanted with such an unusual history?

Meanwhile a request to prepare a chapter, "Control of Development by the Incompatibility System in *Basidiomycetes*" for a book entitled *Secondary Metabolism and Differentiation in Fungi* sat in my inbox. It posed my first challenge to write an updated review of the research Red had spearheaded with his many associates over a period of two decades. I wanted to do my best. I devoted a good part of the fall to this task. At some point the writing came easily.

I recall a sunny afternoon sitting out on our tree-shaded patio in Lexington zipping along with pen to paper, pausing to wonder at the derivation of such inspirational energy. I seemed to be pulling it off *without* Karl Esser's help! An idle thought of Red in spirit guiding my hand could not be ignored.

My first experience in teaching a full-blown course was as visiting lecturer in microbiology at Bridgewater State College just south of Boston in the spring semester of 1979. Despite a dearth of stellar students, I did enjoy the challenge.

My memory retains the antics of one student in particular: While correcting an exam, I could interpret the handwriting of all but one. The exception appeared as a series of indecipherable scribbles. In passing out the graded papers next day, I withheld the errant student's exam and asked to see him after class. "I simply cannot read your handwriting," said I. "Will you please read it for me?" He turned a reddish hue, shook his head and reluctantly confessed, "It's just a bunch of bullshit." Recalling that I did likewise on a second-grade spelling test I had not studied for, some 45 years earlier, I felt rather sorry for him—a bit of empathy. My teacher marked me zero and made me sit all day in class with chewing gum stuck to my nose. I gave my student a zero, without sticking gum to his nose.

All in all I thought I had done a pretty good job as adjunct instructor at Bridgewater and should be seriously considered for an advertised tenure-track position there. Alas this proved to be my first shock as job applicant. An interviewing Dean expressed concern for my interest in doing research, when no current member of the Biology Department did any research at all. Furthermore, he dismissed with a smirk my contention that parenthood provided learning experiences relevant to classroom teaching: *don't drone on in lecture when students' eyes glaze over—try another tack.* Finally, while holding his chin in clasped hands, he asked, "Why would anyone with a Harvard

PhD want to join Bridgewater's Department of Biology in the first place?"

His comment took me by surprise. I had thought this Dean would be pleased to have someone with my credentials who *wanted* to do research. Perhaps he thought my presence would threaten the existing faculty of an institution that had transitioned quite recently from a teachers college to a state college. Or maybe he did not want to provide lab facilities for me. Conceivably, he just did not like me. As I was to learn later, failed applicants seldom know why they are rejected.

I wrote off this nonproductive interview as a learning experience, took my $1,100 for the teaching I had done, and retreated to home base. Subsequent applications to 10 or 12 advertised positions landed three interviews. Finally, Wellesley College, within 20 minutes' drive from home, became my only option. They offered a three-quarter-time instructorship in biology to help team-teach the introductory courses in genetics and developmental biology—it was not the full-time tenure track position I'd hoped for, but better than nothing.

Chapter Twenty-nine

Soaring and Crashing

WELLESLEY IS THE SORT of college I rejected as a student. While scholastically super, its all-female student body and lack of graduate programs registered negatively in my mind. From the day of my interview as prospective faculty member, I sensed a kind of stuffiness, a touch of smugness about the "Wellesley way of doing things" that persisted throughout my five years there. The feeling reminded me of sorority days at Syracuse. I felt rather out of place at the start.

Determined to make the best of the only job offered, I looked on the bright side: after all, it came with an office and a nice little private lab within a brand new building. With outside iron walkways against a windowed facade, the science center stands in stark contrast to the surrounding classical structures on that beautifully foliated campus. The building puts me in mind of the controversial Pompidou Center in Paris, France. Elevated crosswalks span a huge three-story atrium separating offices on one side from labs on the other. Faculty needing to be in both places throughout the day often get waylaid by roving students wanting attention—an advantage for students but a time-consuming bite out of the day for faculty needing to juggle both teaching and research.

Trying to impart enthusiasm for my favorite subject, genetics, to a hundred or more sleep-deprived students, most of whom took the course as a prerequisite for medical school, proved far more difficult than I had imagined. My efforts to use a combination of teaching techniques I thought appealing—slides, chalkboard, overhead projector, toy props, even a participatory dance illustrating chromosome crossover—were met with skepticism. The established methods relied primarily on overhead projectors with lights down low and faces invisible. Only a minority of the students responded positively to my occasional departure from text to point out the vagaries of science. Most found it frustrating and a waste of assiduous note-taking to hear explanations of how ill-conceived experiments might produce useful results, and how the results of well-planned experiments can be misinterpreted even by Nobel Prize winners. Such diversions help little in passing the MCAT for entry into med school.

My status as a neophyte teacher was all too apparent. I didn't even know how to skillfully parlay an incomplete mastery of the subject without faltering. Genetics is principally a problem-solving subject, one in which logical conclusions drawn from quantitative measurements must be understood.

While explaining, for example, the gender-biased inheritance of a recessive sex-linked trait such as hemophilia, it is necessary to work through the pedigree from parents to offspring. The defective gene responsible for the disease is carried on the X chromosome. It cannot encode a necessary step in blood clotting. The disease is most often expressed in male offspring because males have only one X chromosome, which comes from the mother; females have two, one from the mother and one from the father. If the defective version of the gene is carried on one of the mother's Xs (as in the case of Queen Victoria) it has a 50-50 chance of being passed on to all of her offspring regardless of sex. If by chance it is passed on to her son, he will express the disease. If, on the other hand, it is passed on to her daughter, she will not have the disease because the functional, dominant, wild type version of the gene carried on her other X chromosome from Dad complements. One good version suffices for normal function.

To my dismay, no matter how well I thought I knew such material I found it sometimes possible to mess up teaching it before a demanding audience—especially while following several generations of sex-linked traits in the model genetic organism, the fruit fly. I admitted my mistakes. Seasoned teachers know how to avoid such pitfalls; they might pretend they messed up on purpose, just to test the students' degree of alertness. It would take some time before I became seasoned.

The course in developmental biology during second semester went better. There I could wax enthusiastic about developmental systems in fungi, while the other two team-teachers covered plant and animal systems. Most of the students in that course took it by choice, and the atmosphere seemed less rigid. The venue was enviable for its good facilities, ample supplies, and simpatico teammates. I enjoyed introducing eager young minds to a whole kingdom of organisms they had hardly known existed.

The lab portion excited a sense of experimentation: I had the students choose from a variety of fungi to run fungal growth races from the point of inoculation on slants of medium in elongated test tubes and see how controlled variants such as light and temperature affect growth and circadian rhythms. (This exercise was inspired by Jay Dunlap, the grad student who helped teach the micro labs I directed back at Harvard.) They liked to bet on the winner.

They also got a kick out of slicing up slugs of creepy crawly cellular slime molds and tracing regeneration of the head versus hind bits into differentiated stalk and spore cells, as Ken Raper first demonstrated. Remembering Cees van Niel's approach, I knew such play could elicit meaningful discussions about the fundamentals of cell differentiation. These Wellesley gals, interested in science to begin with, responded more positively to an element of discovery in lab sessions than those Harvard non-science students I had tried to reach in George Wald's course.

Happily, a number of able students wanted to try out research in my own lab for credit. Fortuitously, the work with *Schizophyllum* could be managed in small segments as individual projects, and funding was obtainable. Together we kept things perking at a merry hum. I loved mentoring these bright young students from freshman to senior year, watching them grow from bumbling neophytes to budding scientists, learning to question authority and come up with their own ideas. While the biology faculty contained as many males as females, I sensed that these young female undergrads appreciated having a female role model, something my generation seldom encountered.

Most, it seemed, came to Wellesley because their parents deemed it a safe haven, cloistered at some distance from the seaminess of such places as Harvard Square. Entering students tended to shyness. As they developed a Wellesley-nurtured sense of assertiveness in the absence of male competitors—aided and abetted by the emerging feminist movement—these young women began to realize a sometimes overwhelming wealth of opportunities. The atmosphere clearly differed from the Wellesley of two generations earlier when my high school teachers, the "Barkae" studied there.

My research during this period centered on efforts to determine the control of synchronous nuclear division in our organism. Just how nuclei are regulated during their cell cycle is of fundamental importance to all living creatures. It is central to proper development and clearly relevant to development gone awry, as in cancer. *Schizophyllum,* with its precise regulation of synchronous nuclear division, is an obvious system for studying this phenomenon. We knew we could disrupt synchrony by mutation and we knew that the mating-type genes are key players, but which ones and how? Could they regulate nuclear division at some particular phase of the cell cycle from one DNA replication to the next?

Once two nuclei of fully compatible types for both the *A* and *B* mating-type genes get together after a mating, they cozy up together and stay coupled. I watched them, using special luminescent staining techniques. They look like a pair of automobile headlights, one pair in each cell of the mycelium. They always divide at the same time and get distributed as pairs into new adjacent cells where they again divide simultaneously.

I thought communicating signals between the nuclei might control that synchrony. Now that I could isolate those nuclei into separate cells as protoplasts and monitor their rate of division in regenerated protoplasts (as Wessels taught me to do), I could try separating different pairings of known mating-type genes, both mutated and wild type, and see whether they had anything to do with sending or receiving such signals. My students and I set about the task.

To our surprise, we found that the *B* mating-type genes alone affect the timing of nuclear division. By inference, properly functioning *B* genes produce signals for synchrony—if one nucleus reaches the right stage of the cycle for division, it

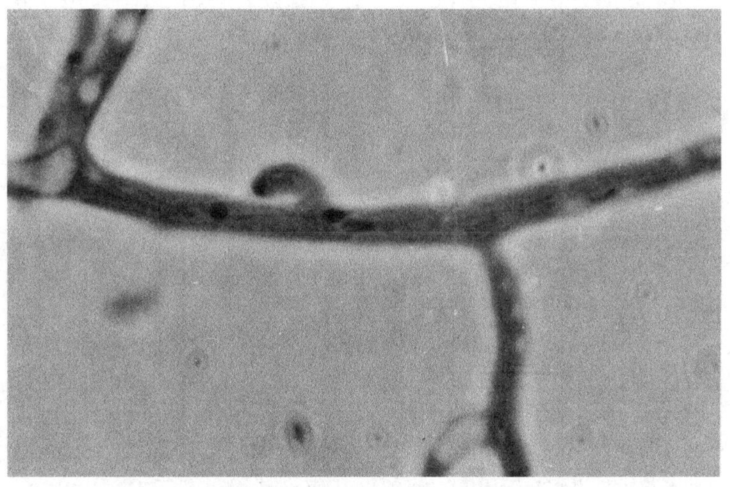

Synchronous nuclear division in Schizo.

waits for the other to catch up, and they can divide together. Interesting, but what is the nature of those signals and how are they received? Once again, the reality of our need for molecular genetic tools became all too apparent. To go further, I would have to learn to master those techniques.

Though I had not yet mastered the comfort and skill of teaching a required subject such as Introductory Genetics, Wellesley did promote me to a full-time tenure-track position as Assistant Professor of Biology after my first year's experience. By then the scene seemed brighter once I'd befriended other junior faculty and had made the acquaintance of cool students, who clearly wanted to learn what I was able to teach. I began to see the professor-student relationship as a two-way street: I imparted knowledge and understanding to students while they rejuvenated my thought processes with fresh ideas I might not have thought of otherwise. We had fun learning together.

Furthermore, invitations to student parties enriched my understanding of student social life. At one such gathering we played a game called "Tree": "If you could be a tree, what kind would you be?" Others decided for you; then you would either agree or disagree. They thought I would be an oak tree; I agreed. It fit my slow-growing, late-blooming, steadfast nature.

One Halloween potluck party stands out in memory: I brought innocent brownies, but someone (probably the student dressed as an atom bomb) showed up with a bowl full of "weed" accompanied by a side of cigarette wrappers. While flattered that the organizers thought me tolerant enough to be included in their circles, I did not partake of the weed—a beer sufficed. I had to drive home.

One of my fellow Assistant Professors, Nigel West, had a keen interest in flying. Flying intrigued me as well. The two of us happened to be gazing at the same notice on a bulletin board in the Science Center atrium one day: "MIT Soaring Association Recruiting New Members—Orientation Meeting." It stated date, place, and time—sometime in January. Nigel looked at me and said, "You wanta go?" I said, "Why not?" We went together for an indoor session and got hooked. A demo ride, including a chance to handle the controls for a brief period after launch, cinched our interest. Thus began a three-year effort to learn how to fly gliders under the tutelage of expert pilots. It consumed many weekends of blessed relief from all the trials and tribulations of trying to be the perfect professor. I had skied, sailed, bicycled, hiked, and played tennis, but nothing could take me so far away as soaring. A spell of Transcendental Meditation helped get me through grad school; soaring helped get me through the demanding years at Wellesley. The tests required for a pilot's license—practical, oral, and written—seemed almost more daunting than requirements for the PhD.

Then came the crash!

That fateful prelude to tenure decision (permanent professorship) began in the fall of 1982. Having assembled what is known as "the package" (list of publications, summary of research and funding, student evaluations, letters of recommendation from outside colleagues, service to the college, professional activities in the world at large), I awaited judgment.

Denial of tenure at Wellesley hit in the spring of 1983. Less than stellar evaluations from students required to take the introductory course in genetics killed me. Those ratings, mandatory just prior to final exams, ranged from "poor" to "very good" and a few "excellents"—but averaged "not good enough." Evaluations in other courses, including Developmental Biology and Horticulture, extended from "okay" to "very good," and ranking for a smaller course in Advanced Genetics scored "very good" to "excellent."

My pattern of success in teaching seemed to correlate with decreasing class size and increasing student motivation for the subject taught. As far as I know, I seem to have passed muster in other areas of endeavor, especially research, but the bottom line was failure, a failure to be shared by three of my favorite junior professors at a comparable level, all axed for different reasons not entirely clear to any of us. We made good company, and stayed good friends. In defiance, we staged a celebratory reunion of *refusés* at my summer cottage two decades later.

With busted ego I started making plans: applications to other institutions netted three interviews, none of which progressed to job offers. The future looked bleak. Would I have to be a dime-store clerk after all?

Clearly, the only hope of continuing in science was mastery of the then-new molecular techniques in application to my research. A proposal to the National Science Foundation (NSF) to spend a summer learning such techniques under the sponsorship of colleagues working with *Schizophyllum* at the University of Vermont did succeed. Bob Ullrich, Associate Professor of Botany at UVM and one of Red's last graduate students (the big fellow Celia falsely accused of putting her behind the eight ball just before her finals), had recently pioneered work on the characterization of *Schizophyllum's* genomic DNA, a necessary prelude to "gene jockeying" in this organism. My postdoctoral advisor, Professor Jos Wessels, and his associates in the Netherlands were also laying the necessary groundwork for studying *Schizophyllum* at the molecular level at about this time.

The summer of 1983 marked the beginning of my entry into this whole new field of genetics.

Chapter Thirty

Wellesley Farewell

In the last two semesters at Wellesley I faced problems requiring resolution: Besides efforts to find another position, I had to devote a great deal of time to completing research, writing up results, and applying for grants—two grants, in fact.

The past summer of research at the University of Vermont, supported by a small supplement to Bob Ullrich's NSF grant, afforded me possible entry to an appointment as Research Assistant Professor in the Department of Medical Microbiology at UVM if I could come up with funding. Bob in the Botany Department collaborated with Charlie Novotny in the smaller Department of Medical Microbiology to clone genes from *Schizophyllum.* I worked in both labs throughout the summer and became acquainted with most members of both departments. By summer's end, the Department of Medical Microbiology, through supportive efforts of its chair Warren Schaeffer, expressed willingness to let me join them in a non tenure-track position to be supported by outside grants.

Bob and Charlie worked diligently that summer to adapt known methods for gene cloning in the yeast, *Saccharomyces cerevisae,* to our mushroom fungus *Schizophyllum*—so far without success. While we had accomplished little in the way of results, I began to learn molecular methods and wanted to know more. The idea of doing research full-time and living in beautiful Burlington, Vermont, not far from my Allen family relatives in Plattsburgh and Montreal, appealed to me.

Another place to start might be in Bill Timberlake's lab at the University of California, Davis. Bill had previously worked on the water mold *Achlya* that Red had pioneered in his earlier years. Now he had moved to another fungus, an ascomycete called *Aspergillus nidulans,* for which he was close to developing a protocol for cloning genes relevant to development. He ran a lively lab, and the research involved a variety of molecular techniques I was keen to learn.

I had taken the initiative to visit Bill earlier in the year while visiting son Jonathan at Stanford, where he was then doing a postdoctoral in developmental neurobiology. Upon our meeting, Bill, a type A person in body and mind, sat nervously beside me on a campus bench drenched in California sunshine. Pumping his hands across his tightly jean-clad thighs and bobbing his curly head, he said, "Well, I suppose you want to come to my lab and work on *Aspergillus.*" "No," I replied, "I really want to come to your lab and work on *Schizophyllum.*" My bravado surprised him.

Unlike John Bonner, who insisted Red work as postdoc on his bean seedling project at Cal Tech, Bill heard me out and did agree to accept my presence in his crowded lab for one or two terms. We would work together in a joint attempt to identify gene transcripts (transcripts from DNA to RNA on the way to making proteins) that are unique to the mating process in *Schizophyllum.* It would, of course, be necessary for me to secure the needed funding.

The application process to the National Science Foundation proved daunting. Typewriters were becoming passé; I had to start learning computer technology. Wellesley had an early system of computer terminals connected to a cumbersome processor called Deckstar at MIT, a system I never quite mastered.

I first tackled the task of applying for a one-year grant from the National Science Foundation's Women in Science Program, designed to promote the advancement of women on their way to prominence in any field of science. The guidelines of this relatively new initiative were clearly specified. Thinking advancement in the realm of research would be appropriate, I proposed to learn skills of molecular genetics applicable to my organism in Bill Timberlake's lab.

Pending an answer to my NSF Women's Program proposal, I started staying up way past bedtime working on a more comprehensive NSF grant to study gene transcripts unique to mating. These would theoretically represent genes that we in the Raper lab had mutated to cause disruption of the sexual process. The central challenge was to learn as much as possible about the relevant molecular techniques before having a chance to use them. Learning through reading is one way to learn. Learning by doing is better. But writing a proposal to do something I had not yet learned how to do posed a significant challenge.

I had scheduled a final draft of the big 15-page proposal to NSF for a Sunday evening, the night before the finished document had to be express-mailed to my sponsoring institution, the University of Vermont, for necessary signatures of approval.

I'll never forget the scene: alone at midnight in the computer room. I edit up through page 13. Exhausted, I need a break. I punch the save buttons, then trudge upstairs to the vending machine for an energizing cup of coffee. Not a soul around, no sounds beyond the constant drone of air-circulating machines. Somewhat revived, I drag myself back to the computer and try calling up the saved document: *No, no, no, it cannot be!* Adrenalin rising, I punch the keyboard harder and harder, over and over. That damn unedited version from page 1 on comes up every single time! Now, too late to call for help, I have a choice: kick up my heels, throw a tantrum, and give up, or start the editing *all over again.* I scream and curse to a room full of dumb computer screens, and, for a third night of sleeplessness, find an extra ounce of spunk to start the grim task of final edits, all over again.

Two of my faithful students, who knew more about computer antics than I, must have sensed impending disaster. They came to the rescue uncommonly early on Monday morning and managed the printing of all fifteen copies of the final draft, while I frantically attended to the bureaucratic necessity of filling out numerous attachments in longhand to be typed up by a sacrificing Departmental secretary who devoted her day to my needs.

I reached the post office just before closing. While carrying me home along Route 128 amidst the bustle of rush hour traffic, my trusty orange Volvo somehow exited one ramp too soon. With a blink of the eyes and shake of the head, I regained control and managed to wend my way back home to sleep away that night on into the next day.

UNWELCOME MAIL ARRIVED EARLY in the spring of 1983. My Women in Science proposal had failed. The reasons given seemed somewhat puzzling: While the evaluating panel deemed the research proposal fine, it faulted me for lack of a formal teaching component; I had not proposed to teach a course at Davis on how a woman might succeed in science. Well after all, I had not yet quite succeeded myself! The panel argued that funded candidates should serve as role models for younger women aspiring to a career in science, and that classroom teaching should be part of the package—a guideline stated nowhere in the program's application forms. Knowledge in hindsight helped not at all. In any case, I wanted to focus then on research, not teaching.

One sunny morning late in spring, the phone rang. I was munching on a ham sandwich at my desk. DeLill Nasser from NSF headquarters came on the line. DeLill, an uncommonly savvy and caring official, headed the program section on Eukaryotic Genetics, to which my big proposal had been sent for scrutiny. She started by saying, "Your proposal has been evaluated and found to lie within the gray area for funding." *Pause—a bite of sandwich stuck in my throat.* She then went on to say, "But my program could manage half the requested funding; the other half might possibly come from another source." *I swallowed with some relief.*

The other source she had in mind was a higher NSF appointee (the highest yet for a woman) who had just been terminated with six months notice by the Reagan administration. This official fortunately favored botanical research, deeming it relatively neglected as compared to research on animal systems. She therefore decided to use her remaining discretionary funds to support research on plants. DeLill suggested we present fungi as part of the plant world. She rather favored the fungi, having previously taken a stimulating course in mycology taught by my brother-in-law Ken Raper in the Department of Botany at the University of Wisconsin. I agreed. Fungal courses had been offered traditionally under the heading, Cryptogrammic Botany, and teaching about fungi was still included within botany curricula more often than not.

DeLill went to work and managed within the week to secure the full package. I came to know her well in subsequent years, greeting her at meetings and serving on the panel she managed. What a blessing she was to the field of genetics, what a buttress for women in science at a time when women sorely needed that encouragement.

Sadly, DeLill chain-smoked. I recall a vision of this tall, dark, curly-headed woman propped against an outside doorway, sucking-in, blowing-out, grabbing merciful relief from duties of the day. She died prematurely of lung cancer. The Genetics Society of America fittingly memorialized her by setting up a fund in her name to support young graduate and postdoctoral trainees in their career advancement. I shall always remember DeLill for giving me, as she did so many others, a chance to start on a new road to discovery, an uncertain venture leading to long-term success.

Funding for three successive years cinched my invitation to be a research-track professor at the University of Vermont, putting me into the next phase of my career as molecular geneticist. I had no problem arranging a plan to spend the fall semester at Davis learning techniques in Bill Timberlake's lab before settling in at UVM in January.

WELLESLEY'S DEPARTMENT OF BIOLOGY bade me farewell by throwing a party at the Chairperson's home. As since noted by others who have coped with departure through denial of tenure, the gathering, while well intentioned, proved somewhat awkward.

I shamefully added to the awkwardness of the affair by my spontaneous negative reaction to a thoughtful gift presented by the Departmental Chair and hostess. She knew I enjoyed the thrills of primitive outdoor camping, especially on the undeveloped 40 acres of wilderness at Peabody Pond in Maine that I'd bought at a bargain price from a friend two years earlier. The Chair purchased a Coleman Lantern from Lechmere Sales in Cambridge. Unbeknownst to her, I harbored a longstanding loathing for the cold brilliance of Coleman Lanterns that pierce the peacefulness of a crackling campfire. Throughout our camping experiences together, both Red and I deplored the unnatural hissing of propane lanterns as much as the clamorous sounds of gas-powered electric generators.

As I loosened the gift-wrapping and read the label at the top of the box I blurted without thinking, "Oh! I *hate* these things!" Silence prevailed, and then a kind of embarrassed chuckling ensued. The Chair, recovering from her astonishment, stammered, "Oh, I was told you needed this at your place in Maine—but if you don't want it you may exchange it at Lechmere."

Fearing my apology did not quite make amends, I did nonetheless exchange the Lantern for a blow-up rubber din-

Basking in Wellesley's farewell gift

ghy in which I could lie naked and paddle around—skinny-dipping being the preferred activity at that remote site on Peabody Pond. One of my young colleagues, who enjoyed camping with me, took a snapshot of my sun-drenched naked bod blissfully bobbing about in the little rubber boat. The photo mysteriously appeared as an uncaptioned posting on the Departmental bulletin board at Wellesley some weeks after my departure.

MEANWHILE, AN INVITATION TO chair a session and present a paper on *Agaricus* at the International Mycological Congress in Tokyo had me planning for an exciting summer of travel. Brother Luther and my friend Hilde Grey expressed interest in coming along, not only to Japan but also on a preceding ten-day tour of eastern China—Beijing, Shanghai, and places in between. The tour group constituted of a congenial busload of curious mycologists with accompanying friends and family. As visiting scientists, we garnered special treatment with formalized tours of microbiological institutes, hospitals, factories, and even a commune specializing in pigs and mushrooms.

Our traveling group in China reflected a diversity of perspectives: Besides scientists, it included politicians, a psychoanalyst (my friend Hilde Grey), social scientists, high school teachers, and stay-at-home moms—a sufficient assortment of viewpoints to pose a variety of penetrating questions to the authorities, who lectured us during institutional visits about population control, advances in child care, women's liberation, and more. Our tour guide programmed almost everything right down to the standard table settings of plum wine and green tea.

Toward the end, members of our group began inquiring more deeply about how the government fostered its touted reforms such as birth control to limit population growth. The answer was female rather than male sterilization. "Why a preference for the more complex procedure of tubal ligations in females over the simpler process of vasectomies in males?" we asked. "Vasectomies cause too many bad side effects," said they.

Government efforts to liberate women seemed fraught with inequities. For the most part women still did the shopping, cooking, and homemaking as well as work in the factories. While the factories did provide daycare, women had to retire five years earlier than men, ostensibly to stay home and take care of the grandchildren.

White-clad police seemed to monitor every picture-taking move. The primitiveness of Chinese society with its bicycle-clogged streets, old buildings, sidewalk food vendors, and modest accommodations in the best of hotels, stood in stark contrast to our subsequent experience in Japan, where neon-lit enterprises and computer-equipped tourist facilities appeared more advanced than comparable venues back home. Tokyo had made big modernizing strides in the 16 years since Red and I visited there for the International Genetics Congress in 1967.

Red had been the one presenting a paper at that Congress. Now I carried on by not only giving a talk but also acting as chair for one of the sessions. I felt proud to have come this far, yet knowing my final goals lay some distance away. All went well enough to be followed up by an invitation to participate comparably at an international meeting of the British Mycological Society at the University of Manchester in England the following spring.

UPON OUR RETURN TO Boston, we managed a jolly family gathering at a rented seaside cottage on Cape Ann. (I had leased the Lexington house just before the trip to China and Japan.)

Linda arrived from her new dwelling in rural Vermont where she had moved and started a new career, not as a teacher, but as a quiltmaker.

Jonathan had since married a fellow graduate student, Susannah Chang. Their doctorates completed, he and Susannah stopped on their way from Stanford (where Jonathan did postdoctoral study) to Tübingen, Germany,

Daughter Linda, quiltmaker

Susannah and son Jonathan, newly doctored

where they were to begin a new research project in neuroscience together at the Max Planck Institute for Developmental Biology.

I was headed for Davis, California, and my new calling as a molecular geneticist in Bill Timberlake's lab.

We celebrated the start of three new ventures—all headed in different directions.

Chapter Thirty-one

Molecular Transition

As in the Netherlands, my stay in Davis was car-less. I had shipped my three-speed Schwinn for transportation, knowing that bicycling suited that flat and sunny place.

A friendly welcome from Bill Timberlake and his students put me immediately at ease. They lost little time in showing me the ropes. Bill would have me meet with him at 6:00 AM to demonstrate the methods I wanted to learn: DNA isolation, RNA isolation, radioactive labeling, fraction

Learning molecular techniques

collecting of messenger RNA transcripts—the whole bit I needed for doing the planned experiments.

Those early morning hours, literally for the birds, merited the sacrifice for such valuable one-on-one instruction. Bill patiently explained the rationale behind each step of each protocol while I took copious notes. If something went wrong in my hands, he helped troubleshoot. His grad students, postdocs, and techs—seven, counting a visiting professor from Holland—all jammed into one big lab space. They did what they could to help me fill in the blanks of necessary knowledge.

This fall at Davis marked my transition from classical to molecular geneticist, a switch that not many at my age were willing or able to make.

Bill's lab reminded me of Red's but with half the numbers in one-sixth the space. Good spirits prevailed most of the time. Friendships developed. Sharp-faced Bill, tidily dressed, generally serious, expected the most from his students. His earnest manner posed an easy target for mocking frivolity—with all due respect. Bill habitually holed up in his office, door closed, from early morning until noon when, by the clock, he appeared in the lab *after* his lunch but just *before* ours, demanding the latest results as our gastric juices flowed.

In response to such predictability, his students and techs hid the results of their current experiments, claiming they weren't quite complete. In truth they often were, but the folks who produced them preferred to have first crack at analysis—one of the perks of doing research—before giving their boss the pleasure.

Unbeknownst to Bill, various members of his group sometimes altered his precisely decreed protocols, either by design or by inadvertent mess-ups. They adapted those changes that made no difference or streamlined the process without informing Bill until, perhaps, some later date.

Despite the ambiance, I began to realize that the difference between doing molecular genetics and plain old classical

Mendelian genetics was somewhat comparable to the difference between cooking by recipe versus cooking by instinct. I honestly preferred the latter but knew I *had* to do the former.

While busy learning, I did enjoy this new California lifestyle—bicycling each day to and from work, renting cars from Rent-a-Wreck to go soaring out of Conestoga or to visit Raper relatives in Santa Cruz. Jaunts to Napa Valley vineyards with members of the Department of Microbiology became special treats. The visited vintners felt so indebted to that Department for its shared scientific expertise in viticulture they brought out their finest for tasting, while discussing just how and under what circumstances each wine was produced. It surprised me to learn, for example, that the esteemed flavor of certain rare dessert wines came from rotting grapes, fortuitously infected by a fungus called *Botrytis*.

Before leaving Davis, I managed to achieve enough results for a short paper describing a transcript we had found to be unique to sexual development—my first official molecular publication.

Taking my cue from the Dutch, I threw my own going-away party early in December. Bill arrived early all dressed properly in creased trousers, jacket, tie—the first I had seen him so formally decked out. Then members of the lab group began drifting in wearing their usual sloppy attire of hung-out shirts over worn-out jeans. Without warning, Bill disappeared for half an hour, only to reappear in a sport shirt and jeans just in time for martinis, beer, pizza, salad, and ice cream. His confidence as a scientist did not extend to a mastery of social skills.

My experience in Bill's lab marked an important transition from classical to modern biology. I had made new and lasting friendships in those three months. It was time to say goodbye to sunny California and start my own lab in wintry Burlington, Vermont.

I SETTLED IN A subleased condominium about a mile from campus. While departmental members welcomed me kindly, the weather did not. That winter, one of the coldest on record, attacked not only me but my Volvo, a vehicle supposedly built to withstand the worst of Swedish weather. One dark night after leaving the lab, I could not get the damned thing started, not even a kick. I trudged to the highway through wind and snow and waved my arm for a lift. No one stopped. Driver after driver just whizzed on by with hardly a glance my way. Left with an overwhelming sense of abandonment, a feeling of rejection from my chosen long-term dwelling place, I walked the twelve blocks home with a whimper and ice-cold feet, longing for California weather.

Another problem arose at the outset: Upon arrival in the assigned lab I found it partially occupied by someone else. Clutter—stuff that had not been touched for years—filled the unoccupied portion. Clean up took weeks, delaying my efforts to find a technician and get established with all the equipment I'd brought from Wellesley and Harvard.

Just after clearing and settling the lab, nausea and fever attacked. I awoke the next morning in a hospital bed, missing a rudely ruptured appendix. Having been uncommonly healthy all my life, this bout set me back a bit, diminishing my smugness.

But quick recovery seemed imperative. That invited talk on the edible mushroom at the British Mycological Society would come up within three weeks. I had the airplane ticket, but had not yet prepared my presentation. The importance of this talk and my attendance at a subsequent meeting in the Netherlands could not be ignored; I gave it a try.

All went well enough at the British meeting—the appendix stitching stayed put throughout my talk and the laughing and joking between sessions—but an evening of gaiety at the subsequent Dutch meeting nearly undid me. When I foolishly joined others in a bout of bowling, some stitches gave way. A good night's rest had a healing effect. I journeyed as planned to rendezvous over Easter with son Jonathan and his wife Susannah at the Max Planck Institute in Tübingen, Germany.

We celebrated Easter Sunday, sitting on their sunny balcony, helping David, a student friend of theirs, build a giant papier-mâché *Drosophila* egg to surreptitiously stuff in the back of his mentor's Volkswagen Bug as an Easter surprise. The mentor, Dr. Nusslein-Volhard, held a senior position at the Institute studying the genetics of developmental processes in the fruit fly *Drosophila melanogaster.* A Nobel Prize rewarded her hard work.

Frau Doktor Nusslein-Volhard did not achieve such distinction by taking time off, even on Easter day. David gleefully jammed the giant egg into the back seat of Nusslein-Volhard's parked Bug and hid behind a bush to observe her reaction as she emerged from the lab after a full day's work: nothing, not even a smile. Apparently unamused, she unceremoniously removed the egg, tossed it aside, and drove on home without looking back. *Some people are too damn serious! Or was she savvy about the prank and just determined not to respond?*

By then my condition had improved sufficiently for me to suggest, "Hey, let's go to Italy." Jon and Susannah, not expecting a Nobel Prize in any case, agreed to the plan. Bright weather prevailed. We drove through Bavaria and Switzerland to Milan and Verona, the famed city of Shakespeare's *Romeo and Juliet,* where we reveled in delightful walks through picturesque neighborhoods and leisurely suppers at outdoor cafes in the sun of April evenings. All in all it was a relaxing vacation that fully restored my sense of well-being before I headed back home and got down to productive work in Vermont.

Chapter Thirty-two

Vermont Landing

I BEGAN TO NOTICE something wrong with what I had thought was a close, long-standing friendship with Bob Ullrich just before the trip abroad to England. Bob was one of Red's last graduate students. He and his wife Sonja socialized frequently with us, and a warm relationship continued even after Red's death. Bob was the student I chose to speak at Red's memorial service. He gave a moving testimony to Red's integrity as a mentor and scientist. I recall him saying something like this:

> . . . In this age of highly competitive science, Red Raper was singular—a stern task-master, but never bullying. Precision, effort and creativity were expected and required. Despite this strictness, students were given a completely free hand to develop their ideas. Whether the thesis topic was closely related to his own work or not was somewhat immaterial.

Bob himself had worked on a fungal system different from *Schizophyllum.*

I thought Bob, as mentor, might follow in Red's footsteps. He had welcomed me as a visiting professor the previous summer, but now as his colleague, I sensed something different—a veiled resentment at my presence. For one thing, he appeared to begrudge my contributions at joint lab meetings. I perceived some reluctance to share thoughts freely. I knew he and my postdoctoral advisor Jos Wessels were rivals, and their disagreements became more apparent when they faced one another at the British meeting we all attended. I wondered if that had anything to do with the chill. All was not as I had hoped. More disappointment would follow.

Back in my lab after the trip abroad, things somehow got organized. I managed to hire as tech a recent college graduate on his hopeful way to med school; we started experiments searching for more transcripts unique to expression of the sexual cycle in *Schizo.*

I could not expect to attract graduate students until I had established a visible presence at the University of Vermont. Inquiries from prospective grad students interested in fungal research naturally came first to Bob and Charlie, and they would not share students with me; why should they? No mandatory system of lab rotations for first-year graduate students either in Botany or in my Department of Medical Microbiology existed at the time, and I was the new kid on the block.

Nevertheless, students in both Bob and Charlie's labs liked to come and talk to me about their work and mine. A free exchange of ideas came easily with them, and I loved it. They seemed eager to learn what I knew about the history of our research with *Schizophyllum* and about fungi in general. In fact, they asked me to teach a summer course in Experimental Mycology. I was pleased to do so, even though the effort would be voluntary without pay. Somehow Bob did not share my enthusiasm for sharing thoughts with *his* students.

Bob's safety valve blew in the presence of my first postdoctoral student Steve Horton when we met with Bob to consider my research proposal to the National Institutes of Health (NIH) in the fall of 1990. I had met Steve three years earlier at a mycological meeting at the University of Toronto while sitting beneath a big apple tree, martini in hand at the end of a day full of scientific talks. Sipping a beer, he eased his trim frame down and greeted me in a friendly manner, "Hi, Dr. Raper, I've always wanted to meet you. I really liked your talk on *Schizophyllum.* I'm completing a doctoral thesis on the molecular genetics of *Achlya* here in Toronto, and am looking for a postdoctoral. Is there a chance I could come to your lab for that?" We prattled on about his research and mine.

I liked what he said. I also liked his quizzical smile, dark-haired good looks, and stylish Canadian attire. We agreed on the spot that he should come to Vermont and do his postdoctoral training with me. We planned initially to clone a gene so closely linked to the *B* mating-type loci that it might serve as a stepping-stone to the *B* genes themselves.

But I'm getting ahead of chronology.

The story of continued success in our efforts to explore the molecular mysteries of mating in *Schizophyllum* dates back to a former graduate student of the ill-fated Mexican Celia Dubovoy. Celia, despite severe physical handicaps, had obtained her doctorate under Red Raper at Harvard in 1973. Fortuitously, at my suggestion, Celia's former graduate student, Alfredo Munoz Rivas, began postdoctoral studies with Bob Ullrich soon after I arrived in Vermont. And thereby hangs a tale:

I had met Alfredo at his behest after visiting my son Jonathan, who then was completing his doctoral program in neurobiology at the University of California, San Diego. Brother Luther and I planned a ten-day tour from San Diego to Mexico, starting with Mexico City. Alfredo met us at the air terminal, took us to our reserved rooms at a charming Spanish-style hotel, and hosted us about that vast, fascinating metropolis. Now, after Celia's untimely death, he was trying to complete his thesis on the identification of genes for making the amino acid tryptophan in *Schizophyllum.* He beseeched me to plead with his departmental chairman, Professor Herrera, that he be allowed to continue his partially completed project on *Schizophyllum,* even though it did not mesh with the Department's declared mission of supporting projects clearly relevant to *Homo sapiens.*

I met with Herrera in our hotel lobby and endeavored to persuade him that a continued study of the complicated tryptophan pathway in *Schizophyllum commune* would contribute to the understanding of an important basic biological phenomenon. He spoke no English; I spoke no Spanish nor the Mexican equivalent thereof. I didn't view Alfredo's research as truly exciting in any case. Either my feigned enthusiasm for the project or my prestige as John Raper's wife did the trick. Herrera allowed Alfredo to go on with the work he had started as Celia's student; in the end Alfredo came up with some mutations in the tryptophan pathway that the Red Raper lab never found.

How did such a discovery impinge on continued funding for *Schizophyllum commune* research? Bob and Charlie had been trying diligently to introduce selectable genes from the bacterium *Escherichia coli* into *Schizophyllum* without success. Transfer of genes from one organism to another is an essential requirement for gene cloning and the molecular manipulation of any organism. *E. coli* was chosen as a tool for its ability to make many copies of specific genes, which could be extracted and integrated into the genome of another organism.

The trick is to find a gene that works in both organisms—in our case, *Schizophyllum* and *E. coli.* Such a gene from *E. coli,* for instance, would be identified as being able to complement a mutated version of that same gene in *Schizophyllum.*

Bob and Charlie had tried several *E. coli* genes that encoded detectable products such as adenine and uracil, but had not tried tryptophan, because a tryptophan mutant was not available—until Alfredo arrived with his mutant strains. Lo and behold, one of them worked! A wild type tryptophan gene of *E. coli* happened to complement one of Alfredo's mutants. For the first time, transformation of cloned DNA into our organism was shown to be possible. Alfredo saved the day.

Success happened in the nick of time, just as I was applying for renewed funding from the National Science Foundation during the end of my second year in Vermont. I was on the phone with DeLill Nasser discussing my proposal when she heaved a sigh, asking "*When* are Ullrich and Novotny going to get transformation in *Schizophyllum*? We just can't keep on funding *Schizophyllum* much longer without some promise of success!" I assured her of a breakthrough; the process would be up and running very soon.

And so it was, thanks to an all-important ingredient furnished through the thread of connection between Red, Celia, Alfredo, my fortuitous talk with Alfredo's departmental chairman after Celia's death, and, of course, the persistent skills of Bob Ullrich, Charlie Novotny, and their students, not to mention Wessels and his students' contributions.

A COUPLE OF YEARS later, Steve and I wrote a grant to the National Institute of Health (NIH) proposing to use the newly developed transformation protocol for the cloning and characterization of a fertilization gene closely linked to the *B* beta mating-type locus. We thought the gene itself would be interesting; and furthermore, if we could snag that gene, we might be able to use it as a bridge to the *B* mating-type genes themselves—a goal long coveted.

We asked Bob and Charlie to critique our proposal before submitting it for funding to NIH. Bob called a meeting in a conference room near his lab a few days later. Upon entering the appointed room, Steve and I sensed much more than critical evaluation as the mood of the moment. In fact it felt like walking into the proverbial lion's den.

Bob sat opposite Steve and me; Charlie, arms folded across his chest, positioned himself a bit to the side. Then Bob, nervously adjusting his half-lens reading glasses, shuffled our proposal in front of him. He started to flush from his neck to his large balding head while expressing an anger that seemed to encompass the entire mass of his formidable body. I'd never seen him in such a state.

I cannot recall the precise words spoken, but the gist of this confrontation was, "You have betrayed our trust. I thought you understood; Charlie and I are going after the mating-type genes. *We* developed the techniques of cloning in our labs, and *we* have first dibs. I thought it was understood that *you* would go after the genes downstream of the mating-type genes, but your proposal looks like you are trying to clone the *B* genes." Steve and I sat, astonished that they assumed priority over research on all the genes in the 59 versions of the four mating-type loci in an organism developed as a model system by many predecessors—including me.

Charlie, "the mountain man," as students often called him for his rugged good looks and colorful cussin' stubbornness, sat stoically by, saying little. I do recall, however, his use of a metaphor to extend Bob's point. The statement went something like this: "There's only one pie to cut into pieces. We want one piece. You can have the rest." But the piece he claimed for himself and Bob had multiple parts of primary interest. That complex, delicious piece had been the key focus of the Rapers' preceding research over a long time. Bob and Charlie had started working on this system only after Red's death. They seemed to think that no one else should even taste it. In retrospect, I suppose they figured funding for such a project could not be stretched beyond their two labs.

The session dissolved without apologies. Steve may have said, "I'm sorry you guys feel that way." He and I walked away shaking our heads, trying to maintain a surface cool to mask deep distress. We spent most of that working day sitting outside on a campus bench, trying to make some sense of it all. I thought back to my childhood as an only girl, the youngest of six children, and how my Mom helped me deal with those five bigger brothers when they tried to lord it over me. I learned then not to give in, lest I become a wimpy, acquiescing female.

Well, now was a time to stand up to two males with academic seniority, not older, but bigger than I.

Steve and I mulled it over. In the end we resolved to pursue our own plans with the conviction that no one had the right to tell us what we were allowed to do in a field of research that had engaged me long before Ullrich and Novotny appeared on the scene. Although others might see it their way instead of our way, we went ahead anyway and submitted our proposal. The award came through four months later.

Needless to say, that incident ended all sense of true friendship I'd ever had with Bob. We greeted one another civilly when meeting, but a wall remained between us. I never went to his house thereafter; he never came to mine. Charlie did stop by my lab the following morning and offered an apology for Bob's loss of control. Fortunately, he and I remained collegial until a later date when competition reared its ugly head once more.

I learned a hard lesson in science firsthand: Research funding had become so competitive it could destroy the advantages of friendly cooperation.

Chapter Thirty-three

Science Trumps Marriage

ONCE THE MOLECULAR TOOLBOX opened for *Schizophyllum,* progress ensued at a rapid pace. The ability to extract, manipulate, and amplify genes for reintroduction into our organism lifted a huge barrier to further discovery. It was now possible to determine directly the phenotype of a single gene. As anticipated, the Ullrich/Novotny teams went for an *A* mating-type gene by walking the chromosome from a closely linked metabolic marker gene. They first cloned the marker gene, then snatched the whole *A* alpha mating-type locus by walking through contiguous clones of all the intervening DNA.

And what did they find? A locus containing not one but two genes with hallmark motifs for making molecules called homeodomain proteins, known to regulate a host of developmental pathways in many other organisms, from insects and amphibians to rats and humans. It was a first. They were riding high.

Others had found comparable genes in distantly related fungi with only two sexes, but this was the first time such genes had been found as part of a system sporting thousands. With two *A* loci, one specifying 9 different mating types, the other specifying 32 different mating types (288 all together) and each locus containing at least two such genes, the entire *Schizophyllum* species must encode a family of at least 82 different versions of homeodomain proteins just to help run sex! Variety may be the essence of life among species as Darwin noted in his *The Origin of Species,* but this is variety *within* a species, designed to promote sexual promiscuity and the consequent intermingling of all the genomes therein.

In *Schizophyllum* any two members of the molecules encoded by the *A* genes must be different in very precise ways in order to fit as partners (to mate, as it were) and turn on a part of the sexual pathway to reproduction. How in the world did nature contrive to generate so many locks and keys, pegs and holes, to run such a system?

And that's only the half of it. Wait 'til we find out what the *B* genes do to complete the other half!

ALTHOUGH STEVE AND I failed in our search for that other half, we found a gene we weren't even looking for. We captured a transformant capable of making mushrooms all by itself in the absence of the normal requirement for compatible mating. Now that grabbed our interest! *It might also interest the edible mushroom industry,* we thought, *even though* Schizo *is not truly gourmet.* We cloned the gene and dubbed it *frt,* for "fruiter." The *frt* gene not only induces mushrooms in unmated strains, it also enhances fruiting in compatible matings.

Our publication on this gene in the journal *Genetics* caught the attention of the biotech corporation known as Genencor in South San Francisco. They had an interest in gene manipulations that might prove profitable to the mushroom-producing industry. They thought it possible that *frt* could induce super fruiting if integrated within the genomes of related crop-producing mushrooms.

Steve and I attended several meetings with Genencor scientists. The meetings progressed to the point of having their lawyers draft a patent application for *frt.* Then a business manager arrived on the scene. Balding, amply built, properly attired in a suit and tie with pockets full of chalk, he pro-

Enhanced Schizo *mushroom production with genetically introduced* frt *gene*

claimed acceptable timelines for the various parts of our plan while drawing arrowed lines connecting labeled boxes across the full length of a chalkboard. "A protocol for putting your gene into the crop mushroom must be accomplished within *x* months," he declared. "Your gene must be hooked up to the appropriate DNA sequence to work in the crop mushroom within *x* more months. It should be shown to enhance mushroom production within the year."

Science being fraught with risks, we could not feign assurance that our timeline would fit his. Heck, a protocol for putting DNA into the crop mushroom had not even been developed, and we knew from experience how long that could take. Without such assurance, the whole plan got squelched as "not cost effective."

So much for a brief glance at possible riches from royalties in support of continued research! Nonetheless, the process itself provided insight into the corporate world and their way of thinking. We cherished retreat to the somewhat more forgiving, more flexible world of academia.

STEVE HAD BEEN WITH me three years when another able student, Lisa Vaillancourt, applied for a postdoctoral position in my lab. A graduate student from Purdue University working with a fungal pathogen of corn crops, she bore credentials good enough to win fellowships in support of her training for a total of three years in my lab. With two able postdocs, a new graduate student Kevin Laddison, and a sizable grant from NIH in hand, I thought us ready to soar in search of the *B* mating-type genes themselves.

The feeling of buoyancy deflated early, however. Steve and Lisa began to annoy each other; they just did not get along. Steve had reached a point of seniority that indicated his readiness to leave my lab and establish one of his own—a status attesting not only to his abilities as a teachable student but to mine as an effective mentor. He knew how to do the things he wanted to do and judged Lisa's questioning manner as uncomfortably confrontational.

Lisa, an emerging feminist with career ambitions, thought Steve unreasonable. I wondered if Steve's resentment might have something to do with gender, but knowing the woman he married, I think not. He often deferred to his wife's wishes, sometimes against his own, such as when she wanted to keep and breed three or four Clumber Spaniels requiring long walks daily.

We sought help through the University counseling services. All three of us attended two sessions together; then the counselor asked to see Steve and Lisa separately. Meanwhile our whole department moved from its original quarters to a brand new building called Stafford Hall, where I had two separate labs. Remembering Red Raper's solution to such problems, I put Steve in one lab and Lisa in the other.

The combination of counseling and separation helped effect a tolerable truce until Steve left for a tenure-track position at Union College, Schenectady, New York. He continued research on *frt* and ultimately climbed the ladder to tenured professorship. Lisa and Steve became pretty good friends after that.

I FOUND SOLACE IN a new relationship of my own. Living in Shelburne at the time, I happened to browse through the local freebee newspaper one day when I spotted an ad in the personal column: "Widower, Williams College graduate, ex-army officer, seeking companionship with woman who loves sailing and classical music"—or something to that effect. What the heck, the Williams College bit implied this widower had some smarts, and a combination of sailing and classical music held definite attraction. I wasn't so sure about the ex-army part. I rang the number given and liked the answering voice. John Cole and I agreed to meet for dinner at a popular watering spot called the Sirloin Saloon, just south of Burlington on Shelburne Road.

It was the first and only time in my life that I sought a date through the media.

I found John Cole there ahead of me, an omen indicating army attitudes. Nonetheless, I liked what I saw: a lean, tall, devilishly handsome fellow a few years older than I. While sipping martinis at the bar, awaiting our table for dinner, we found a mutual link to the past. John's father (John Cole I) had served as the Methodist minister in my hometown of Plattsburgh, New York, when I was but a child. While too young to remember his sermons, I dimly recalled the name from an Allen family anecdote. My middle brother Frank, when herded off to church, complained: "Mr. Cole—I'll *shubble* him!"

Thus began a warm relationship lasting about four years during the early 1990s. John lived in Tucson, Arizona, and served as a volunteer ranger at Organ Pipe National Park in the cooler part of the year. He summered in a rustically charming lakeside cottage built by his grandfather on Thompson's

Point just south of Shelburne. Just offshore he moored his 26-foot cruiser called *Rondo,* so named by his recently deceased wife for its tendency to go in circles when John first learned to sail.

I too owned a little sailboat, *Welcome,* moored at my camp by the lake on North Hero, one of the bigger islands north of Burlington in Lake Champlain. (After moving to Vermont, I had exchanged the place in Maine for a summer cottage closer by.) With what time I could spare from work, we spent summers at his place or mine, sailing each other's boats, investigating islands up and down the mountain-lined lake, attending outdoor classical concerts, meeting for dinners and movies, and generally enjoying each other's company.

John's family welcomed me (somewhat), my family welcomed John (sort of); we started thinking of marriage.

The whole relationship ended, however, with an inability for either of us to make major compromises. Either I had to quit my science and join John in Arizona during half the year, or John had to give up his work in Arizona and live with me in Burlington. John rejected the concept of staying put and relating long distance over part of the year. He needed a living-together marriage. I could not give in, having moved beyond Pappenheimer's old-school axiom about the futility of training women in science: "The woman would just follow her man wherever he had to be."

Alas, John's love for me did not exceed his love for Arizona, and my love for him did not exceed my love for science. I was reminded of three powerhouse female scientists I know, each of whom achieved full professorship, membership in honorary scientific societies, authorship of numerous books and scientific papers, the bearing of four or five offspring—and marriage ending in divorce with two or more husbands. I'd asked these superwomen, "How did you manage so many accomplishments?" Each replied, "Well, the husbands had to go."

I WAS RIGHT NOT to give up on science. The most exciting discoveries lay ahead. While Steve and I had not yet succeeded in cloning the *B* mating-type genes using our experimental approach, Ullrich's former student, Charlie Specht, phoned from Boston one day with the astonishing news that he had cloned some of the *B* genes in his spare time as a postdoc on another project in Phil Robbins' lab at MIT. Not quite ready to give up on *Schizophyllum*, Charlie devised a novel scheme—one *we* should have thought of—to snag these genes from an appropriate clone bank he and fellow graduate student, Luc Giasson, had made before leaving Bob Ullrich's lab. Charlie was devoting full-time to a different project in Robbins' lab at that point. He therefore offered the *B* clones to me for further study.

"Why did you not call Bob and offer them to him first?" I asked.

Specht's answer conveyed resentment: "Bob knows I cloned these genes from a library I made in his lab, but he hasn't asked for them."

I wondered how Bob might have known.

I honestly don't know who thought what when. Why had those two not communicated directly? I do know that Specht had left Bob's lab in something of a huff.

Now he insisted, "I want *you* to have the clones." Well, how could I refuse? I thanked him for his confidence in me and asked, "May I share them with collaborators?" He answered, "You can do what you want." Feeling somewhat chagrined over our inability to clone those genes in the first place, I felt not only delighted but also flattered to be chosen to carry on the work with them.

Accepting Specht's offer and knowing that a *B* gene project would involve a lot of hard work, I started seeking collaboration with other scientists also interested in deciphering the mating system of *Schizophyllum*. I might have asked Ullrich and Novotny to participate, but they were deep into all those *A* genes. Recognizing Specht's reticence and remembering

Bob's antipathy towards me, the idea of sharing the *B* gene project with them remained a passing thought.

Now Novotny got upset. He thought it dishonorable of me to accept Specht's clones. What to do?

Using Specht's clones, Lisa probed one of our available DNA libraries and found a clone encompassing the entire *B* beta locus. She agreed to share that clone (not Specht's sans permission) with Novotny who assigned it to a new student for use in isolating and analyzing the *B* locus of another mating type we were not then working on. Unfortunately nothing new came of it, and Novotny stayed unfriendly. Expressing no further interest, he gave up on *B* and continued concentrating on *A*.

Another collegial relationship lost to competitiveness. Nonetheless, all of our mutual students stayed friendly. Good for them; they saw no need to perpetrate rivalries.

Chapter Thirty-four

Pheromones and Thousands of Sexes

UPON RECEIPT OF CHARLIE Specht's clones, our lab took on a whole new aura. Knowing there are 81 different versions of *B* mating-types specified by the two linked *B* loci, alpha and beta, we wanted to get some idea of the number and nature of the genes involved. As Red had said, "They *have* to be interesting."

It would be a long and arduous journey. I invited two European colleagues to join in the venture: Marjatta Raudaskovsy at Helsinki University, Finland (a former postdoc in Red Raper's lab), and Erika Kothe at Philipps University, Marburg, Germany (a former postdoc in Bob Ullrich's lab). They responded with positive enthusiasm. With agreement as to who would concentrate on what, we shared Charlie's treasure with our new collaborators.

Erika and her students came in first with a sequenced gene of the *B* alpha locus.

To our amazement and delight, they discovered that the gene encodes a receptor protein commonly used for sensing throughout the animal kingdom—yet further support for evolutionary connections between fungi and animals. The receptor protein threads itself seven times through the surface membrane of cells to transmit signals from outside the cell to inside the cell. Such receptors with comparable motifs belong to a large family of molecules found throughout a variety of organisms, from other fungi to insects and mammals. They act as communicators, switch boxes turned on by signals from outside cells to regulate a complex pathway of developmental events within cells.

Erika planned to publish her initial finding, but we cautioned that the receptor gene would likely be just the tip of the iceberg, one component of several to be found within the *B* alpha locus. There had to be signaling genes nearby. She agreed to hold off and wait for the bigger story yet to come.

We used Specht's clones as stepping-stones to larger chunks of DNA, all to be sequenced and tested for function. None of us enjoyed the tedious process of DNA sequencing back in the early 1990s—juggling huge glass plates, pouring bubble-free gels to sequence thousands of DNA components. It was fussy business punctuated by cusswords. We made thankful use of an expertly run modern sequencing facility as soon as it became possible, but the earlier efforts paid off.

Sure enough, in addition to the receptor gene we found several smaller genes all jammed together so closely within one piece of DNA that they separate by meiotic recombination only rarely. Each of the smaller genes encodes small peptides known as pheromones that serve as signals to activate a receptor.

This was just the beginning. We looked at other *B* loci. All are built similarly: one receptor and several pheromone genes, as many as six or seven in some cases. A lot of redundancy permeates the system. Since a single interaction is sufficient to initiate the fertilization pathway, a need for more seems superfluous. Such redundancy may reflect a kind of fail-safe mechanism—nature's way of backing up an essential function in case something goes wrong, as scientists have found for many essential processes in humans (two kidneys for instance).

We used some of the mutations of the *B* loci that Eva Schneider and I had generated two decades earlier to figure out how the pheromones and receptors work.

As molecules, the pheromones act rather like males while the receptors act more like females. Unlike *Achlya,* where the female hormone initiates dalliance, the male-like pheromones initiate sexual interplay in *Schizo.* A pheromone comes along and tries to titillate the receptor. The receptor, like a lock on a gate, embraces the pheromone. If the fit is just right, the gate opens and passes a message along the track to fertilization. Because compatible mates in nature normally express both pheromones and receptors, they can titillate one another; this process of fertilization is reciprocal.

Knowledge at the molecular level verified at least part of my original hypothesis derived from mutational evidence: The *B* mating-type locus contains two kinds of genes, one for donating and one for accepting fertilizing nuclei. I hadn't thought about the two coming together to switch things on. But that was something Red had thought possible decades earlier—I thought it ingenious at the time, and so it was.

As with the *A* gene products, the *B* genes encode a large series of molecules that fit precisely in pairs to sense a proper mate.

Molecules similar to those encoded by the *B* genes are used for sensing a variety of things such as sight, smell, and taste throughout the animal kingdom. Mother Nature, having contrived them, seemed reluctant to let them go but modified and used similar versions over and over to accomplish other tasks throughout the course of evolution.

We were the first to discover a multiple system of pheromones and pheromone receptors governing compatibility among a large number of potential mating partners. Why is this noteworthy? Now it is possible to mutate, mix, and match

Mating type is determined by four loci

A-complex B-complex

Aα Aβ Bα Bβ

9 versions 32 versions 9 versions 9 versions

9 x 32 x 9 x 9 > 20,000 possible mating types !

Differences determine mating compatibility

Multiple mating types from multiple genes with multiple variants

Mating between two haploid strains

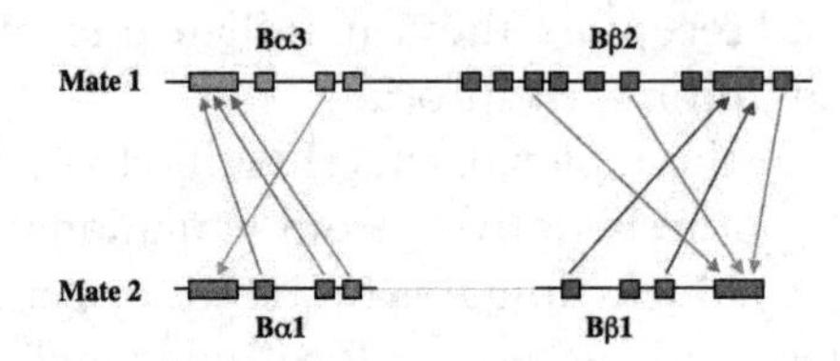

Enormous redundancy for signal activation
No cross-talk between Bα and Bβ

Multiple signals activate a compatible mating

the genes involved and get some idea as to how their products actually fit together to accomplish their mission—a venture not so feasible in other known systems.

OUR JOINT PUBLICATION OF this work marked a milestone: It initiated a final triumph in my lifelong quest to discover something new and important. One colleague wrote to me with over-the-top congratulations: "Cardy, you've found the Holy Grail!"

I was pleased also to accomplish another goal: effective mentoring. My graduate student Kevin Laddison finished his thesis and went on to a good-paying job with Pfizer Pharmaceutical; postdoc Lisa Vaillancourt departed, having landed a tenure-track assistant professorship at the University of Kentucky where she later advanced to tenure. Meanwhile our European collaborators worked on other aspects of the mating process in *Schizo.*

My subsequent postdoctoral student Tom Fowler, technician Mike Mitton, and I continued with studies of other genes within both *B* alpha and *B* beta loci. It had now become possible to collect all the marbles and play with them in all combinations. We cavalierly called our effort to find all the signals and sensors for mating in *Schizo* the *Kama Sutra* project. In truth, with so many to find, such ambitions would require an enormous amount of work. An example of what we found analyzing the pheromone genes of just one mating-type locus is illustrated below.

Then came the prospect of a brand-new, faster way to test how each gene works. It evolved from a seminar Tom pre-

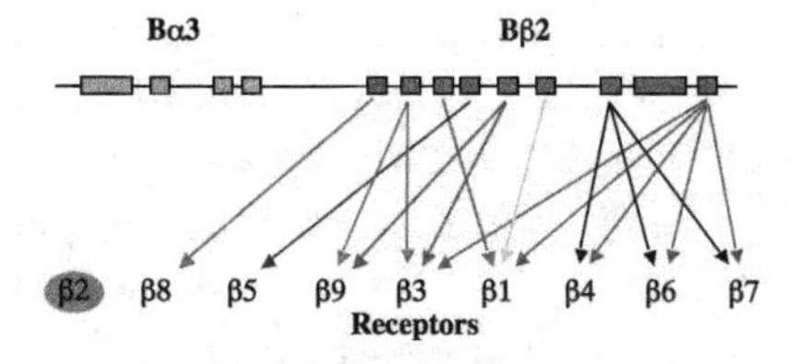

One pheromone can activate several receptors

More than one pheromone can activate the same receptor

All non-self Bβ receptors can be activated

Estimate: ~130 of ~400 possible interactions of Bβ pheromones and receptors result in signaling

Activity of pheromones encoded by genes within a single mating-type locus

sented to a small group of colleagues within our department, a somewhat more formal format than the tea-table talks in Red Raper's lab. After Tom had explained our early results of *Kama Sutra,* an attentive postdoc, Susan DeSimone in Janet Kurjan's yeast lab (across the hall), wondered whether or not our pheromone and receptor genes might function in the more tractable yeast system. I recall Janet's skeptical expression as she shook her red head at such a thought—while acknowledging it wouldn't hurt to try.

Susan went for it. She, Tom, and Mike worked together to put the relevant genes of *Schizophyllum commune* into the appropriate strains of *Saccharomyces cerevisiae,* a very different looking fungus distantly related to *Schizo.*

Upon completion of the first experiment, Susan bounced into our lab with a big grin on her face, shouting, "Hey guys, it worked! It worked!" Hugging each other, we jumped for joy, dropped everything, hopped across campus, and celebrated with ice cream cones all around. Such a feat, opening the door to all the great tricks already known for manipulating genes in yeast, meant far faster progress.

Of course, as every good scientist must, Susan repeated the experiment the next day to verify the first result. *Horrors! Nothing! It simply did not work! Was our ice cream fete premature?*

As every good scientist should, Susan did some troubleshooting. She carefully compared the methods used in each experiment and tracked a difference in the types of pipettes used to harvest the secreted pheromone. She had happened to use glass pipettes in the first attempt but plastic pipettes in the second. When, for the third try, she went back to glass pipettes, the results came out positive.

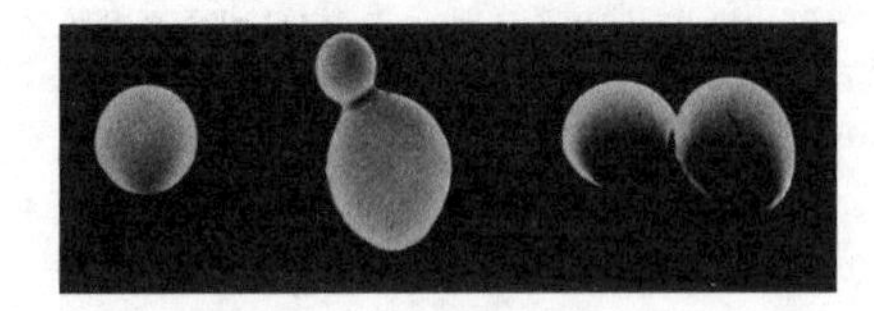

The yeast fungus Saccharomyces cerevisiae

My lab group celebrating the triumph of expressing Schizo B *mating type genes in the yeast Saccharomyces*

Aha: The pheromone molecules, being slimy lipopeptides, stick to plastic but not to glass, thus preventing delivery of the secreted pheromone to the receptor cells being tested. Had Susan used plastic rather than glass in her first try, we might have abandoned the project then and there. We would have had to defer to Janet: the whole idea seemed quite preposterous.

Although yeast is often used to express genes from other unrelated organisms, we deemed our results a marvel.

Astonishingly, it is now possible to make the bisexual yeast multisexual by endowing it with several different compatible combinations of *Schizo* mating-type genes—a fun trick we resisted trying right away, there being more important things to do. In any case the use of yeast to test numerous combinations of natural and mutated pheromone and pheromone receptor genes of *Schizophyllum* vastly expedited our *Kama Sutra* project.

It was at about this time that John Cole and I were contemplating marriage. But I simply could not tear away and follow him to Arizona. The apex of my life as scientist had just been reached. I had been at the University of Vermont for 10 years, and the lab functioned better than ever. Nevertheless, awarded funds from grants were simply not sufficient to cover as much as three-fourths of my salary, buy necessary supplies, and still pay a postdoc and tech. Now, at age 70, I had accumulated sufficient savings for a comfortable retirement. I decided to retire officially but stay active. My chairperson Susan Wallace seemed happy to let me have the office and lab as professor emerita as long as I captured enough money to keep it running.

I never had a big group, but those in it worked productively. Besides Tom (lower left in last previous photo) and Mike (upper left), an assertive visiting scientist from Russia arrived, and my first postdoc Steve Horton returned with an NSF grant for summer research. We had a steady string of undergrads as well. Tom took the lead in hands-on research while I spent more time advising and grant writing, just as Red had had to do.

I couldn't have wished for a better associate. Tom, a modest fellow of quiet demeanor and humble posture, has exceptional intellect, a strong work ethic, and golden hands at the bench. He had come to my lab after a postdoctoral stint in the Plant Science Department with my friend Mary Tierney, where he learned the techniques of molecular biology as applied to plants.

Tom seemed shy at first. He had a painfully hard time talking about his work in public. With considerable tutelage and practice, however, he gained enough confidence to deliver excellent talks at prestigious meetings and ready himself for a tenure-track position at another university when the time came. Everyone liked him, the undergrads especially for his amiable skill in teaching effectively without intimidation.

I admired Tom too for his tolerance of our boyishly handsome and playful tech Mike, who had the incurable habit of listening to lousy music through mandated earphones while humming along out of tune. Mike bided his time as tech, hoping to qualify eventually for med school. That did not happen. Nonetheless, when Mike left to make a more lucrative living in real estate we missed his buoyant spirit and valuable assistance.

As a professor on a research track at the University, I did little formal teaching. However, when Tom and I were asked to give an advanced course on teaching graduate students how to teach, I readily agreed. I had always enjoyed teaching eager students in small groups. We gave our students the option of choosing topics within the general area of our favorite subject, signaling and sensing, including an array of systems, such as sight, odor, taste, and mating in a diversity of organisms, such as mammals, fish, flies, and fungi.

Students had two opportunities to present their chosen material. They had to do some background reading in preparation for each class session and to participate in critical discussion about each selected topic. As part of the course, several professional speakers addressing a variety of relevant systems came as invited presenters at departmental seminars; our students happily took turns hosting the speakers at a free lunch after each seminar. Tom and I agreed: this is a fun way to teach. We encouraged self-criticism and found satisfaction in seeing our students improve as the semester progressed. I think we learned nearly as much from them as they did from us in a mutually enjoyable process.

Meanwhile, our research advanced at a rapid pace. Now that we knew the difference between pheromone and pheromone-receptor gene function, we pondered again the mystery of how so many different mating types could have evolved.

Clues might be found by analyzing at the molecular level those many *B* mutants the Raper lab had generated some two decades earlier.

We got clues all right, but they didn't solve the problem. The process of going from one mating type to another is more complex than we could discern. It must involve several, precise mutations in both types of genes, but we know not how—or why. What is the possible advantage of about 20,000 mating types anyway? Well, it does enhance the rate of genome mixing.

For example, any individual within the species *Schizophyllum commune* is compatible with about 98 percent of all the other individuals it happens to meet within the worldwide population. Sounds like science fiction! Such prolific out-breeding creates a rich variety of different genotypes and a greater opportunity for fitness to a changing environment. Of course such profuse proclivity to propagate consumes energy. The tradeoff for *Schizo* must be worth the cost. The species is abundant; it rots fallen logs and recycles carbon all over the world.

OUR ANALYSES OF THE old mutations did provide clues for intelligently designing new mutations to be tested in yeast, with an eye towards understanding just how pheromones titillate receptors. Our *Kama Sutra* project had entered a speedway toward progress. It continues even now.

Concomitant studies of another related mushroom type fungus, *Coprinus cinereus*, with a comparably elaborate mating complex, were being conducted in Lorna Casselton's lab at Oxford University. She too found a pheromone-receptor based system of communication but with the genes arranged differently, in three loci rather than two. She also detected genes for homeodomain proteins comparable to those found in *Schizophyllum.*

It began to appear that numerous versions of these kinds of genes constitute the basis for multiple sexes throughout the mushroom-bearing fungi. In fact, the molecular genetics of fungal mating was becoming a hot topic, especially since it bore relevance to a variety of sensing mechanisms in animals—including the human kind.

Ironically the field became dominated by what my first postdoc Steve Horton called a bunch of *alpha females.* I had been asked to organize a session on fungal sexuality at a conference on the molecular genetics of fungi, Asilomar, 1999. Female speakers outnumbered males five to one. We females all sat in the front row of the crowded conference room and spoke assertively with exceeding enthusiasm.

I could not help but reflect on the striking reverse in gender roles from the old days when Red and I attended such meetings together.

Chapter Thirty-five

Passing It On

HAVING *OFFICIALLY* RETIRED IN 1994, I decided it was time to *actually* retire ten years later. Now at 80, I had run my own lab for 30 years. Except for studies on the genetics of mushroom production, most of what associates and I did would be classified as basic or curiosity-driven research that might or might not help solve problems confronting mankind.

The mating molecules we discovered are similar enough to molecules regulating development and sensing in humans that our system should aptly serve as a model for understanding how those molecules work in humans. Ironically, son Jonathan identified receptors comparable to those used for mate sensing in fungi as role players in development of the nervous system in zebra fish. They resemble molecules associated with the nervous system in humans as well.

Think how often we say, "We have good chemistry." Could sexual compatibility between humans be based on comparable molecules—I like his; he likes mine? There is evidence for this in rats. Beyond sex, the type of receptor molecules we discovered in *Schizo* are fundamentally involved in human diseases such as multiple sclerosis, diabetes, and drug addiction. We are now in a position to use the fungal system for studying these molecules in ways not possible in more complex species. I find that satisfying.

COMPETITION FOR THE FUNDING of basic research in model organisms has become increasingly fierce over the decades of my belated career. I must have written about 30 grant proposals as an independent scientist—I had to, just to stay in the game. Twenty were awarded; some made it to just this side of gray.

I had gotten the hang of the process, even enjoyed it in some measure. But after a while, begging for funds began to grind me down. The requirements to justify the proposed research and its impact on the entire field of biology, to say nothing of possible benefits to the welfare of humankind, left me feeling not quite honest. How can you know the results of research not yet done? The best, most optimistic thoughts must be cast to sell the proposed work. Contingency plans must be carefully defined in case the preferred approach does not pan out. I've shared dark feelings with my colleagues about the necessity of exaggerating the overall importance of my proposed research in every attempt to wring dollars out of the National Institutes of Health, National Science Foundation, American Cancer Society, United States Department of Agriculture, and other such funding agencies. Rejection wounded my ego; yet the harshest criticism came from colleagues who agreed to review my best drafts before submission. Their tough love, while hard to bear, buffered less harsh critiques from anonymous peers engaged by the agencies themselves. Still, the system works amazingly well; there just never seem to be enough funds for all the good ideas proposed.

Despite frustrations, nothing beats the delight of discovery, the thrill of being the first to know some facet of nature and build upon that knowledge to find out more, then more. It is an endless voyage full of surprises, worth the hard work of devising, performing, and interpreting experiments; managing personnel; writing papers and grant proposals; mentoring, teaching, and attending meetings. Science for me has been an adventure, a pioneering journey, just as my teacher Dr. Rusterholtz first imparted to me back in elementary school.

Preparation to persist and succeed in my profession started when I was very young. Beginning in childhood I was toughened up by five older brothers who tried not only to dominate but also to test me. When we played football, *they* assigned *me* the role of center. When playing Cowboys and Indians, *they* more often were cowboys; *I* was the chased Indian. In "toss the blanket" *they* tossed *me*. I had to prove my worth at every task. I remember one failure in particular: Our parents went out for the evening, leaving me in charge of preparing supper. I cut my finger opening a can of beans. Those miserable brothers laughed, "You can't even open a can of beans!" I threw down the can opener and ran out of the house proclaiming to run away forever. They knew I'd just go to my friend's house across the street. When I tried to sneak back home all the doors were locked. I knocked and yelled. They let me fume just long enough—then let me in, laughing.

Those brothers set the groundwork for my future. Their tough caring helped me meet and triumph over failures and misfortunes. As we all grew up they gained more and more respect for me. "Little Cardy, her head is so hardy" actually went on to do something beyond motherhood and homemaking. She not only finished college but went on to graduate school and became a successful scientist. What's more, she found a good man who egged her on to do almost anything a good man could do.

When my good man left me in mid-life, I might have given up. Had he lived a normal lifespan I would have stayed his wife and partner. As it was, I grasped an opportunity to become more than my husband's wife and reach a scientific breakthrough we might not have been able to achieve together.

I used to think the years in Red's lab at Harvard were the best—I called them the "golden years" in which I could devote full attention to the research and let Red worry about how to keep it all going. Well, now I think the period of my career after his death—the hardest time—was tops. Building upon what he taught and encouraged me to do, I turned my grief to something worthy of our life together. The sense of loss persists even now but not in a debilitating way. I have used it to do something that honored and justified our union. While in younger years I may have dreamed of a knight on white horse, Red was the one who came to my rescue and gave me the necessary sense of pride to accomplish what I did.

The process throughout has nurtured me, 'til my ego approaches that of those Hahvahd professors whom I ridiculed back when I was but the wife of one. Perhaps developing a sense of self-worth that others come to appreciate is one route to happiness. It has been for me.

So HERE I AM, about to close that scientific enterprise. After six years in my lab, postdoc Tom was ready to run his own program. He had reached the point where he could teach me more than I could teach him. We both needed to move on.

My colleagues must have thought so as well when they planned a surprise: By the spring of 2003 I had attended every session of the Fungal Genetics meetings at Asilomar, California, since 1985. On the shores of the wild Pacific within the confines of state park lands set about with tasteful rustic buildings, Asilomar is a favorite meeting place. Scientists come from all over the world to learn of each other's current research. Over two decades, attendance at these biannual conferences grew from about 200 to a capacity crowd of 800. After several days of talks, workshops, and informal exchanges, each conference ends with a festive banquet and a popular talk by some distinguished member. (I had once been so honored.)

On banquet night 2003, Tom surreptitiously maneuvered me to a central table. After dessert but before the special talk,

At Asilomar with former postdocs Lisa, Tom, and Steve

he quietly left the table for the front of the room. Then raucous music started playing. There was Tom, poised at the mic announcing a tribute to Cardy Raper upon her retirement.

I could not believe it; such a dedication to any one member just did not happen at these meetings. Astonishingly, he "roasted" me for the next ten minutes, showing ancient pictures of Red, me, and our favorite fungi all punctuated by amusing anecdotes I didn't know he knew. Glad that Red and I should be so honored for studying sex in fungi, I felt sad that he was not there to enjoy it.

I was called to the podium to receive a commemorative t-shirt and offer a few words of response. After thanking all for such a surprise tribute upon my retirement, the memory of a wild day sailing that could have been my last came to mind. I told the story: "One brisk day last summer my son Jonathan's family and I boarded our sailboat in Burlington Harbor. As we entered the broad lake the wind whipped up, reaching gale force at 35 knots, almost too much for our 22-foot boat with only the jib sail flying. We headed downwind

Son Jonathan at the helm

for my summer place 35 miles north in North Hero, Vermont. The waves lunged so high they engulfed the stern. We sailed alone—no other fools in sight. Jonathan cautioned, 'Put on your life jackets. I can't come about for a rescue if you fall overboard.' Then with a devilish smile he said, 'But Mom, if she goes overboard—she's had a good life!'"

Cardy at 83

Epilogue

I SURVIVED THE DANGER of sailing only to face near death in 2004 when my large intestine ruptured and a skillful surgeon managed rescue on an emergency basis. Conditions failed to improve during the following year. Test after test indicated the dread diagnosis: *cancer.* Oh, I'd rather have drowned!

Miraculously, a second operation encouraged by a second doctor's opinion revealed not cancer but serious inflammation. With health restored by removal of the offending tissues, I have another lease on life—a chance to finish this memoir and stay around to find out more about how our mushroom fungus keeps all its sexes straight.

My return to healthy living also brought two subsequent accolades: 1. Appreciative dedication of a volume called *Sex in Fungi* to my husband and me for our pioneering research in the field, and 2. An all-day symposium in my honor entitled "Themes on Signaling and Sensing" at the University of Vermont in 2008. (I was serenaded at the celebratory banquet by a jazz quartet singing, "How Can We Miss You When You Won't Go Away?")

Following several fungal and fly talks by former students and colleagues, my own son Jonathan capped the session with a talk about his discovery of molecules similar to mushroom pheromone receptors that figure prominently in guiding development of the nervous system. This endearing meeting of mother and son in science made what Jon had called my *good life* even better. Linda—now married to her college sweetheart—came and listened in. Although a colorful quilt artist, she claimed appreciation for all those scientific talks, especially her brother Jon's when, by way of introduction, he uncharacteristically expressed tender sentiments about the positive influences his mother and father had upon him. Beyond all that science, I felt inordinately proud to be the parent of these two.

Former postdocs Tom Fowler and Steve Horton now take the lead in our research: Steve is progressing with work on the genetics of mushroom production at Union College, where he holds a tenured position as teacher par excellence. Tom landed a tenure-track position as Assistant Professor of Biology at Southwestern University, Illinois, in the summer of 2005, and has since progressed to tenure. With several publications to his credit, he works to understand just how a multitude of pheromones—an estimated 100 or more—discriminate among some 18 receptors to activate certain ones and not others.

The entire genome of *Schizophyllum commune* has now been sequenced. It revealed even more sex genes than we thought, plus numerous genes of potential use in making paper out of wood.

Perhaps one day Steve, Tom, their students, and other colleagues will figure out just how the numerous genes devoted to sex in this model fungus operate cooperatively to make a beautiful mushroom.

Mushroom of Schizo

Glossary

allele one of two or more alternative versions of a given gene occupying the identical position (locus) on a chromosome that affects function of a single product (RNA and/or protein).

antheridium the threadlike male sex organ that fertilizes the female sex organ, **oogonium**, in water molds.

basidiomycete a member of the higher fungi that produce macroscopically visible fruiting bodies containing sporulating sex cells called basidia. The most familiar members of this group produce mushrooms—edible, poisonous, hallucinogenic, and otherwise.

chromosome in eukaryotes, one of the threadlike structures within the nucleus comprised of DNA and chromatin and carrying genetic information arranged in a linear sequence.

clone a group of genetically identical cells or organisms propagated from one ancestor, or, in molecular genetics, replication of a DNA fragment placed in an organism.

disomy an unbalanced condition of having one or more chromosomes present in twice the normal number but without having the entire genome being doubled.

DNA deoxyribonucleic acid localized especially in the nuclei of eukaryotes. It is the molecular basis (blueprint) of heredity in most organisms and is constructed in the form of a twisted double strand (a complementing **double helix**) with a polysugar-phosphate backbone held together by hydrogen bonds between specific pairs of bases (adenine to thymine and guanine to cytosine, commonly referred to as A, T, G, and C respectively). Genes contain specific sequences of these four bases in sets of three.

flagella microscopic threadlike appendages of bacteria, protozoa, water molds, spermatozoa, etc. that propel cells by a whipping motion.

fruiting body the sexual spore-producing structure of a fungus, e.g. mushroom.

gene a stretch of DNA that encodes a single product, e.g. protein, that is transmitted from parent to offspring.

genetic locus a position on the chromosome containing a specific gene or genes.

genome the complete hereditary genetic material of a single organism.

genotype the genetic characteristics of an organism eliciting a **phenotype.**

haploid a cell or organism containing a single copy of genetic material on one set of chromosomes, as contrasted to **diploid**, having a duplicate set of chromosomes.

homeodomain protein any of a particular class of proteins that regulates the transcription of genes involved in a developmental pathway of events.

homozygous the condition of having identical allelic forms of a single gene or genes, as contrasted with **heterozygous**, in which the alleles of specific genes differ.

hormone a regulatory molecule that is transported in fluids, usually within but sometimes between organisms to stimulate a new activity in cells distant from those that produce it.

hypha each of the branching filaments that make up the mycelial body of a fungal colony.

mating-type loci specific chromosomal loci containing sets of closely linked mating-type genes that determine the sexual compatibility of appropriate mates. In the mushroom-bearing fungi the A loci contain genes encoding homeodomain proteins; the B loci contain genes encoding pheromones and pheromone receptors.

mitosis the vegetative process of nuclear division of either haploid (usually in fungi) or diploid nuclei, resulting in two daughter nuclei of identical genetic content, as contrasted to the sexual process of **meiosis**, in which diploid nuclei mix the genetic content of their chromosomes through recombination and reassortment to produce four haploid daughter nuclei of differing genetic content.

molecule the smallest unit of a compound that can be involved in a chemical reaction.

morphology the discernible form of an organism.

mutagenic agent any substance that causes mutations in DNA.

mutation a change in the structure of a gene through alteration of its DNA that can be passed from one generation to the next—a **mutant** is an individual carrying the mutation.

mycology the study of fungi.

phenotype the observable characteristics of an individual resulting from its genetic constitution, the **genotype**.

pheromone a molecule released into the environment by an organism, which affects the behavior of another organism (usually of the same species) having the appropriate **pheromone receptor**.

protein any of a class of nitrogenous organic compounds consisting of a long string of amino acids. Proteins make up the structure of body tissues.

protoplast the protoplasm of a living bacterial, fungal, or plant cell which has had its cell wall removed.

RNA ribonucleic acid produced from a DNA template. Messenger RNA is read in codons and serves as a template for the making of specific proteins.

spores differentiated, reproductive single cells of microorganisms, each capable of germinating and developing into the whole organism. **Vegetative spores** are derived through mitosis; **sexual spores** are derived through meiosis.

taxonomy the science of classifying organisms.

transcripts lengths of RNA transcribed from DNA templates.

transformation the genetic alteration of a cell or organism by introduction of extraneous DNA.

wild type a gene, strain, or characteristic that is typical and prevails within a population in nature, as contrasted to an atypical **mutant type**.

About the Author

University of Vermont Research Professor Emerita in Microbiology and Molecular Genetics, Cardy Raper is renowned for her innovative studies of fungi, their function in nature, and how they reproduce. She is best known for her pioneering research on the determinants of sexual reproduction in a ubiquitous wood rotting mushroom with more than 20,000 sexes, in which the controlling molecules resemble those involved in sight, taste, and smell in humans.

Born the only girl with five older brothers, Cardy Raper dreamed of becoming a scientist when girls were not thought suitable for such a profession. Despite lack of encouragement, she traveled a convoluted path to finally establishing herself as an independent scientist with her own laboratory. Her autobiography chronicles that journey, both in and out of the laboratory. She delves deeply into the intertwined dramas of love, parenthood, political activism, and an academic career in science.

The author is extensively published in national and international scientific journals. She has also authored several book chapters. She received a Masters in Science degree from The University of Chicago and a PhD from

Harvard University. She was awarded over 20 grants for her research from the National Science Foundation, the National Institutes of Health, and the United States Department of Agriculture among others. She was an invited speaker at numerous international symposia and at colleges, universities, and industries in the U.S., Canada, Europe, Israel, and Japan. At a retirement dinner and symposium in her honor, scientists came from all over the world. Dr. Raper is listed in "Who's Who of Women" and "Who's Who in Frontiers of Science and Technology" and was recently honored as a dedicatee for the book *Sex in Fungi.* As an Emerita Research Professor in Microbiology and Molecular Genetics at the University of Vermont, she has published her memoir which received an Independent Publishers award for excellence in the category of science, and was recently elected Fellow of the American Academy for the Advancement of Science.

Acknowledgements

THIS MEMOIR would never have reached publication without the expert and caring encouragement of my editor Linda Bland of Cahoots Writing Services. After devoting a career to research science and its publication, I had little understanding of how to write about my life as scientist in a way non-scientists and scientists of other disciplines could appreciate. Linda's continuing guidance through countless revisions proved invalu¬able. I also appreciate the helpful influence of meetings with the League of Vermont Writers, as well as a writing workshop with author Mark Pendergrast.

I am indebted copy editor Jenny Holan and to the friends and relatives who read versions of this work as it progressed, offering constructive criticisms along the way: my son Jonathan; daughter Linda Carlene; daughter-in-law Susannah Chang; nieces Helen Nerska, Carol Allen, and Emma Jean Bowman; grandniece and professional editor Elizabeth Allen; colleague Marjatta Raudaskoski; literary agent Gary Heidt; and friend Charlotte Lichterman (who helped most in weeding out the more obscure scientific terminology). I also acknowledge a helpful critique provided by Annelise Finegan, Acquisitions Editor of Syracuse University Press; she steered me toward telling my story with more of me in it and away from an inclination to write as a less personified scientist. Dede Cummings and Carolyn Kasper of DCDESIGN provided the skillful formatting and design I could never have done myself. I thank my publisher, Hatherleigh Press, for their support: Andrew Flach, along with the team of Ryan Tumanbing, Anna Krusinski, and Ryan Kennedy. Remeline C. Damasco, MD, was gracious enough to write the Foreword for the book.

Finally, I am grateful for the kind words of my reviewers Ursula Goodenough, Peter Day, Faye Crosby, Lorna Casselton, and Willem Lange.